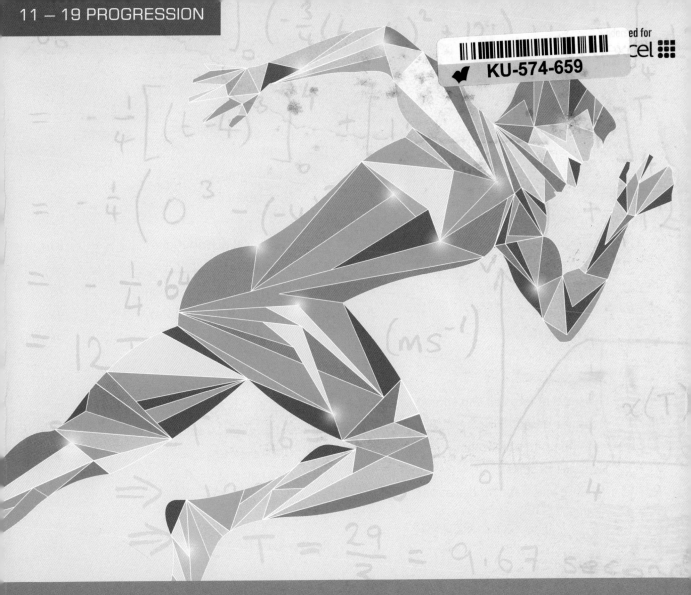

Edexcel AS and A level Mathematics

Statistics and Mechanics
Year 1/AS

Series Editor: Harry Smith
Authors: Greg Attwood, Ian Bettison, Alan Clegg, Gill Dyer, Jane Dyer, Keith Gallick,
Susan Hooker, Michael Jennings, Jean Littlewood, Bronwen Moran, James Nicholson,
Su Nicholson, Laurence Pateman, Keith Pledger, Harry Smith

Pearson

Contents

Overarching themes

The following three overarching themes have been fully integrated throughout the Pearson Edexcel AS and A level Mathematics series, so they can be applied alongside your learning and practice.

1. Mathematical argument, language and proof

- Rigorous and consistent approach throughout
- Notation boxes explain key mathematical language and symbols
- Dedicated sections on mathematical proof explain key principles and strategies
- Opportunities to critique arguments and justify methods

2. Mathematical problem solving

- Hundreds of problem-solving questions, fully integrated into the main exercises
- Problem-solving boxes provide tips and strategies
- Structured and unstructured questions to build confidence
- Challenge boxes provide extra stretch

The Mathematical Problem-solving cycle

specify the problem → collect information → process and represent information → interpret results →

3. Mathematical modelling

- Dedicated modelling sections in relevant topics provide plenty of practice where you need it
- Examples and exercises include qualitative questions that allow you to interpret answers in the context of the model
- Dedicated chapter in Statistics & Mechanics Year 1/AS explains the principles of modelling in mechanics

Finding your way around the book

Access an online digital edition using the code at the front of the book.

Each chapter starts with a list of objectives

The *Prior knowledge check* helps make sure you are ready to start the chapter

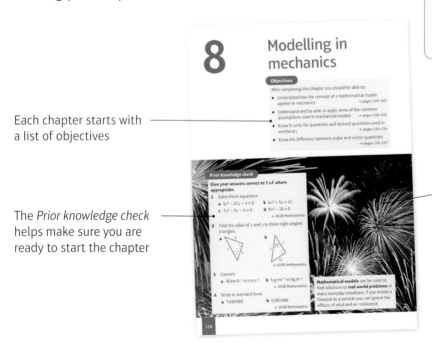

The real world applications of the maths you are about to learn are highlighted at the start of the chapter with links to relevant questions in the chapter

Exercise questions are carefully graded so they increase in difficulty and gradually bring you up to exam standard

Exercises are packed with exam-style questions to ensure you are ready for the exams

Problem-solving boxes provide hints, tips and strategies, and *Watch out* boxes highlight areas where students often lose marks in their exams

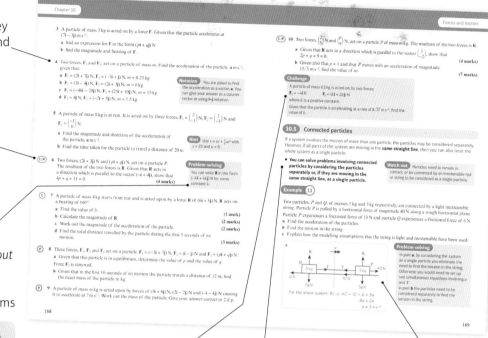

Exam-style questions are flagged with Ⓔ

Problem-solving questions are flagged with Ⓟ

Each chapter ends with a *Mixed exercise* and a *Summary of key points*

Challenge boxes give you a chance to tackle some more difficult questions

Each section begins with explanation and key learning points

Step-by-step worked examples focus on the key types of questions you'll need to tackle

Every few chapters a *Review exercise* helps you consolidate your learning with lots of exam-style questions

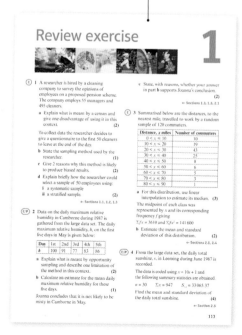

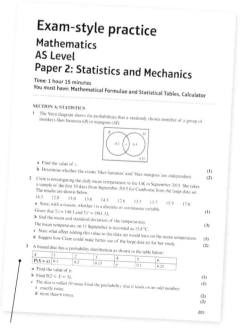

A full AS level practice paper at the back of the book helps you prepare for the real thing

Extra online content

Whenever you see an *Online* box, it means that there is extra online content available to support you.

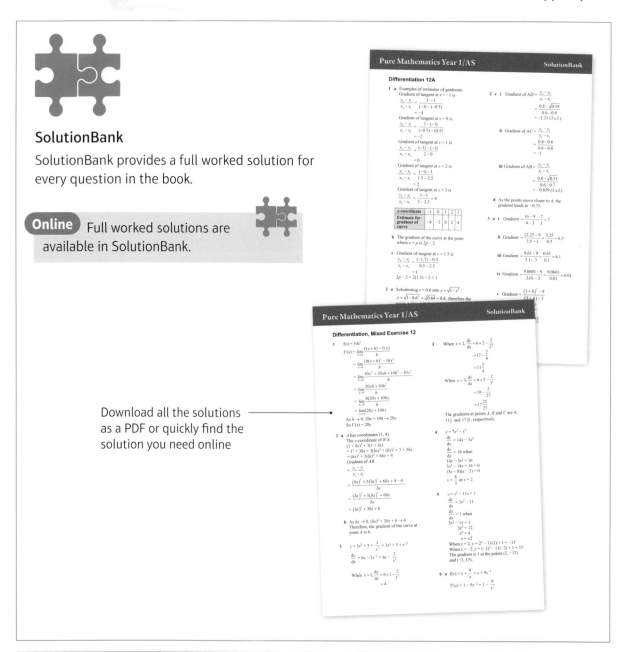

SolutionBank

SolutionBank provides a full worked solution for every question in the book.

Online Full worked solutions are available in SolutionBank.

Download all the solutions as a PDF or quickly find the solution you need online

Access all the extra online content for free at:

www.pearsonschools.co.uk/sm1maths

You can also access the extra online content by scanning this QR code:

Use of technology

Explore topics in more detail, visualise problems and consolidate your understanding. Use pre-made GeoGebra activities or Casio resources for a graphic calculator.

Online Find the point of intersection graphically using technology.

GeoGebra

GeoGebra-powered interactives

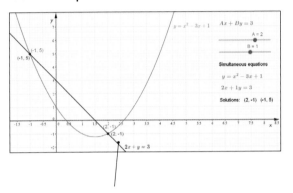

Interact with the maths you are learning using GeoGebra's easy-to-use tools

CASIO®

Graphic calculator interactives

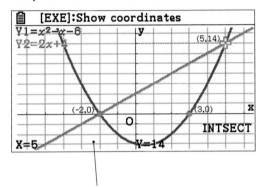

Explore the maths you are learning and gain confidence in using a graphic calculator

Calculator tutorials

Our helpful video tutorials will guide you through how to use your calculator in the exams. They cover both Casio's scientific and colour graphic calculators.

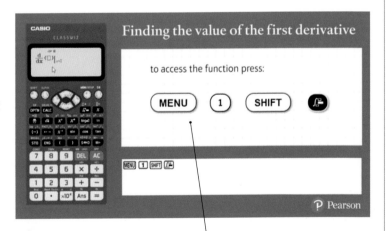

Finding the value of the first derivative

to access the function press:

MENU 1 SHIFT

P Pearson

Online Work out each coefficient quickly using the ^nC_r and power functions on your calculator.

Step-by-step guide with audio instructions on exactly which buttons to press and what should appear on your calculator's screen

Published by Pearson Education Limited, 80 Strand, London WC2R 0RL.

www.pearsonschoolsandfecolleges.co.uk

Copies of official specifications for all Pearson qualifications may be found on the website: qualifications.pearson.com

Text © Pearson Education Limited 2017
Edited by Tech-Set Ltd, Gateshead
Typeset by Tech-Set Ltd, Gateshead
Original illustrations © Pearson Education Limited 2017
Cover illustration Marcus@kja-artists

The rights of Greg Attwood, Ian Bettison, Alan Clegg, Gill Dyer, Jane Dyer, Keith Gallick, Susan Hooker, Michael Jennings, Jean Littlewood, Bronwen Moran, James Nicholson, Su Nicholson, Laurence Pateman, Keith Pledger and Harry Smith to be identified as authors of this work have been asserted by them in accordance with the Copyright, Designs and Patents Act 1988.

First published 2017

20 19 18 17
10 9 8 7 6

British Library Cataloguing in Publication Data
A catalogue record for this book is available from the British Library

ISBN 978 1 292 23253 9

Printed in the UK by Bell and Bain Ltd, Glasgow

Acknowledgements
The authors and publisher would like to thank the following individuals and organisations for permission to reproduce photographs:

(Key: b-bottom; c-centre; l-left; r-right; t-top)

Fotolia.com: Arousa 156, 197cr, aylerein 20, 113tl (a), mdesigner125 1, 113tl, Okea 40, 113cl; **Getty Images:** Billie Weiss / Boston Red Sox 69, 113cr, bortonia 98, 113tr, rickeyre 130, 197cl; **Shutterstock.com:** Anette Holmberg 59, 113c, Carlos E. Santa Maria 118, 197tl, Fer Gregory 181, 197tr, Joggie Botma 123, John Evans 83, 113tr (a)

All other images © Pearson Education

Pearson has robust editorial processes, including answer and fact checks, to ensure the accuracy of the content in this publication, and every effort is made to ensure this publication is free of errors. We are, however, only human, and occasionally errors do occur. Pearson is not liable for any misunderstandings that arise as a result of errors in this publication, but it is our priority to ensure that the content is accurate. If you spot an error, please do contact us at resourcescorrections@pearson.com so we can make sure it is corrected.

Contains public sector information licensed under the Open Government Licence v3.0.

Data collection

1

Objectives

After completing this chapter you should be able to:

* Understand 'population', 'sample' and 'census', and comment on the advantages and disadvantages of each → **pages 2–3**

* Understand the advantages and disadvantages of simple random sampling, systematic sampling, stratified sampling, quota sampling and opportunity sampling → **pages 4–9**

* Define qualitative, quantitative, discrete and continuous data, and understand grouped data → **pages 9–10**

* Understand the large data set and how to collect data from it, identify types of data and calculate simple statistics → **pages 11–16**

Prior knowledge check

1 Find the mean, median, mode and range of these data sets:

   a 1, 3, 4, 4, 6, 7, 8, 9, 11   **b** 20, 18, 17, 20, 14, 23, 19, 16

                                                 ← GCSE Mathematics

2 Here is a question from a questionnaire surveying TV viewing habits.

> How much TV do you watch?
>
> ☐ 0–1 hours   ☐ 1–2 hours   ☐ 3–4 hours

   Give two criticisms of the question and write an improved question.   **← GCSE Mathematics**

3 Rebecca records the shoe size, x, of the female students in her year. The results are given in the table.

   Find:

   a the number of female students who take shoe size 37

   b the shoe size taken by the smallest number of female students

   c the shoe size taken by the greatest number of female students

   d the total number of female students in the year.   **← GCSE Mathematics**

x	Number of students, f
35	3
36	17
37	29
38	34
39	12

Meteorologists collect and analyse weather data to help them predict weather patterns. Selecting weather data from specific dates and places is an example of sampling.

                 → Section 1.5

1.1 Populations and samples

■ **In statistics, a population is the whole set of items that are of interest.**

For example, the population could be the items manufactured by a factory or all the people in a town. Information can be obtained from a population. This is known as raw data.

■ **A census observes or measures every member of a population.**

■ **A sample is a selection of observations taken from a subset of the population which is used to find out information about the population as a whole.**

There are a number of advantages and disadvantages of both a census and a sample.

	Advantages	Disadvantages
Census	• It should give a completely accurate result	• Time consuming and expensive • Cannot be used when the testing process destroys the item • Hard to process large quantity of data
Sample	• Less time consuming and expensive than a census • Fewer people have to respond • Less data to process than in a census	• The data may not be as accurate • The sample may not be large enough to give information about small sub-groups of the population

The size of the sample can affect the validity of any conclusions drawn.
- The size of the sample depends on the required accuracy and available resources.
- Generally, the larger the sample, the more accurate it is, but you will need greater resources.
- If the population is very varied, you need a larger sample than if the population were uniform.
- Different samples can lead to different conclusions due to the natural variation in a population.

■ **Individual units of a population are known as sampling units.**

■ **Often sampling units of a population are individually named or numbered to form a list called a sampling frame.**

Example 1

A supermarket wants to test a delivery of avocados for ripeness by cutting them in half.

a Suggest a reason why the supermarket should not test all the avocados in the delivery.

The supermarket tests a sample of 5 avocados and finds that 4 of them are ripe.
They estimate that 80% of the avocados in the delivery are ripe.

b Suggest one way that the supermarket could improve their estimate.

a Testing all the avocados would mean that there were none left to sell.

When testing a product destroys it, a 'census' is not appropriate.

b They could take a larger sample, for example 10 avocados. This would give a better estimate of the overall proportion of ripe avocados.

In general, larger samples produce more accurate predictions about a population.

Exercise 1A

1 A school uses a census to investigate the dietary requirements of its students.

  a Explain what is meant by a census.

  b Give one advantage and one disadvantage to the school of using a census.

2 A factory makes safety harnesses for climbers and has an order to supply 3000 harnesses. The buyer wishes to know that the load at which the harness breaks exceeds a certain figure.

  a Suggest a reason why a census would not be used for this purpose.

  The factory tests four harnesses and the load for breaking is recorded:

        320 kg   260 kg   240 kg   180 kg

  b The factory claims that the harnesses are safe for loads up to 250 kg. Use the sample data to comment on this claim.

  c Suggest one way in which the company can improve their prediction.

3 A city council wants to know what people think about its recycling centre.
The council decides to carry out a sample survey to learn the opinion of residents.

  a Write down one reason why the council should not take a census.

  b Suggest a suitable sampling frame.

  c Identify the sampling units.

4 A manufacturer of microswitches is testing the reliability of its switches. It uses a special machine to switch them on and off until they break.

  a Give one reason why the manufacturer should use a sample rather than a census.

  The company tests a sample of 10 switches, and obtains the following results:

        23 150   25 071   19 480   22 921   7455

  b The company claims that its switches can be operated an average of 20 000 times without breaking. Use the sample data above to comment on this claim.

  c Suggest one way the company could improve its prediction.

5 A manager of a garage wants to know what their mechanics think about a new pension scheme designed for them. The manager decides to ask all the mechanics in the garage.

  a Describe the population the manager will use.

  b Write down the main advantage in asking all of their mechanics.

1.2 Sampling

In random sampling, every member of the population has an equal chance of being selected. The sample should therefore be **representative** of the population. Random sampling also helps to remove **bias** from a sample.

There are three methods of random sampling:

- Simple random sampling
- Systematic sampling
- Stratified sampling

■ **A simple random sample of size n is one where every sample of size n has an equal chance of being selected.**

To carry out a simple random sample, you need a sampling frame, usually a list of people or things. Each person or thing is allocated a unique number and a selection of these numbers is chosen at random.

There are two methods of choosing the numbers: generating random numbers (using a calculator, computer or random number table) and **lottery** sampling.

In lottery sampling, the members of the sampling frame could be written on tickets and placed into a 'hat'. The required number of tickets would then be drawn out.

Example 2

The 100 members of a yacht club are listed alphabetically in the club's membership book.

The committee wants to select a sample of 12 members to fill in a questionnaire.

a Explain how the committee could use a calculator or random number generator to take a simple random sample of the members.

b Explain how the committee could use a lottery sample to take a simple random sample of the members.

a Allocate a number from 1 to 100 to each member of the yacht club. Use your calculator or a random number generator to generate 12 random numbers between 1 and 100.
Go back to the original population and select the people corresponding to these numbers.

> If your calculator generates a number that has already been selected, ignore that number and generate an extra random number.

b Write all the names of the members on (identical) cards and place them into a hat. Draw out 12 names to make up the sample of members.

■ **In systematic sampling, the required elements are chosen at regular intervals from an ordered list.**

For example, if a sample of size 20 was required from a population of 100, you would take every fifth person since $100 \div 20 = 5$.

The first person to be chosen should be chosen at random. So, for example, if the first person chosen is number 2 in the list, the remaining sample would be persons 7, 12, 17 etc.

- **In stratified sampling, the population is divided into mutually exclusive strata (males and females, for example) and a random sample is taken from each.**

The proportion of each strata sampled should be the same. A simple formula can be used to calculate the number of people we should sample from each stratum:

The number sampled in a stratum = $\dfrac{\text{number in stratum}}{\text{number in population}} \times$ overall sample size

Example 3

A factory manager wants to find out what his workers think about the factory canteen facilities.

The manager decides to give a questionnaire to a sample of 80 workers. It is thought that different age groups will have different opinions.

There are 75 workers between ages 18 and 32.

There are 140 workers between ages 33 and 47.

There are 85 workers between ages 48 and 62.

a Write down the name of the method of sampling the manager should use.

b Explain how he could use this method to select a sample of workers' opinions.

---

a Stratified sampling.

b There are: 75 + 140 + 85 = 300 workers altogether.

$18-32$: $\dfrac{75}{300} \times 80 = 20$ workers.

$33-47$: $\dfrac{140}{300} \times 80 = 37\frac{1}{3} \approx 37$ workers.

$48-62$: $\dfrac{85}{300} \times 80 = 22\frac{2}{3} \approx 23$ workers.

Number the workers in each age group. Use a random number table (or generator) to produce the required quantity of random numbers. Give the questionnaire to the workers corresponding to these numbers.

Find the total number of workers.

For each age group find the number of workers needed for the sample.

Where the required number of workers is not a whole number, round to the nearest whole number.

Each method of random sampling has advantages and disadvantages.

Simple random sampling	
Advantages	**Disadvantages**
• Free of bias • Easy and cheap to implement for small populations and small samples • Each sampling unit has a known and equal chance of selection	• Not suitable when the population size or the sample size is large • A sampling frame is needed

Systematic sampling	
Advantages	**Disadvantages**
• Simple and quick to use • Suitable for large samples and large populations	• A sampling frame is needed • It can introduce bias if the sampling frame is not random

Stratified sampling	
Advantages	**Disadvantages**
• Sample accurately reflects the population structure • Guarantees proportional representation of groups within a population	• Population must be clearly classified into distinct strata • Selection within each stratum suffers from the same disadvantages as simple random sampling

Exercise (1B)

1 a The head teacher of an infant school wishes to take a stratified sample of 20% of the pupils at the school. The school has the following numbers of pupils.

Year 1	Year 2	Year 3
40	60	80

Work out how many pupils in each age group will be in the sample.

b Describe one benefit to the head teacher of using a stratified sample.

Problem-solving

When describing advantages or disadvantages of a particular sampling method, always refer to the context of the question.

2 A survey is carried out on 100 members of the adult population of a city suburb. The population of the suburb is 2000. An alphabetical list of the inhabitants of the suburb is available.

a Explain one limitation of using a systematic sample in this situation.

b Describe a sampling method that would be free of bias for this survey.

3 A gym wants to take a sample of its members. Each member has a 5-digit membership number, and the gym selects every member with a membership number ending 000.

a Is this a systematic sample? Give a reason for your answer.

b Suggest one way of improving the reliability of this sample.

4 A head of sixth form wants to get the opinion of year 12 and year 13 students about the facilities available in the common room. The table shows the numbers of students in each year.

	Year 12	Year 13
Male	70	50
Female	85	75

a Suggest a suitable sampling method that might be used to take a sample of 40 students.

b How many students from each gender in each of the two years should the head of sixth form ask?

5 A factory manager wants to get information about the ways their workers travel to work. There are 480 workers in the factory, and each has a clocking-in number. The numbers go from 1 to 480. Explain how the manager could take a systematic sample of size 30 from these workers.

6 The director of a sports club wants to take a sample of members. The members each have a unique membership number. There are 121 members who play cricket, 145 members who play hockey and 104 members who play squash. No members play more than one sport.

   a Explain how the director could take a simple random sample of 30 members and state one disadvantage of this sampling method.

The director decides to take a stratified sample of 30 members.

   b State one advantage of this method of sampling.

   c Work out the number of members who play each sport that the director should select for the sample.

1.3   Non-random sampling

There are two types of non-random sampling that you need to know:

- Quota sampling   • Opportunity sampling

- **In quota sampling, an interviewer or researcher selects a sample that reflects the characteristics of the whole population.**

The population is divided into groups according to a given characteristic. The size of each group determines the proportion of the sample that should have that characteristic.

As an interviewer, you would meet people, assess their group and then, after interview, allocate them into the appropriate quota.

This continues until all quotas have been filled. If a person refuses to be interviewed or the quota into which they fit is full, then you simply ignore them and move on to the next person.

- **Opportunity sampling consists of taking the sample from people who are available at the time the study is carried out and who fit the criteria you are looking for.**

> **Notation**   Opportunity sampling is sometimes called **convenience sampling**.

This could be the first 20 people you meet outside a supermarket on a Monday morning who are carrying shopping bags, for example.

There are advantages and disadvantages of each type of sampling.

Quota sampling	
Advantages	**Disadvantages**
• Allows a small sample to still be representative of the population • No sampling frame required • Quick, easy and inexpensive • Allows for easy comparison between different groups within a population	• Non-random sampling can introduce bias • Population must be divided into groups, which can be costly or inaccurate • Increasing scope of study increases number of groups, which adds time and expense • Non-responses are not recorded as such

Opportunity sampling	
Advantages	**Disadvantages**
• Easy to carry out • Inexpensive	• Unlikely to provide a representative sample • Highly dependent on individual researcher

Exercise 1C

1 Interviewers in a shopping centre collect information on the spending habits from a total of 40 shoppers.

   a Explain how they could collect the information using:

     i quota sampling   **ii** opportunity sampling

   b Which method is likely to lead to a more representative sample?

2 Describe the similarities and differences between quota sampling and stratified random sampling.

3 An interviewer asks the first 50 people he sees outside a fish and chip shop on a Friday evening about their eating habits.

   a What type of sampling method did he use?

   b Explain why the sampling method may not be representative.

   c Suggest two improvements he could make to his data collection technique.

4 A researcher is collecting data on the radio-listening habits of people in a local town. She asks the first 5 people she sees on Monday morning entering a supermarket. The number of hours per week each person listens is given below:

     4   7   6   8   2

   a Use the sample data to work out a prediction for the average number of hours listened per week for the town as a whole.

   b Describe the sampling method used and comment on the reliability of the data.

   c Suggest two improvements to the method used.

5 In a research study on the masses of wild deer in a particular habitat, scientists catch the first 5 male deer they find and the first 5 female deer they find.

   a What type of sampling method are they using?

   b Give one advantage of this method.

   The masses of the sampled deer are listed below.

Male (kg)	75	80	90	85	82
Female (kg)	67	72	75	68	65

   c Use the sample data to compare the masses of male and female wild deer.

   d Suggest two improvements the scientists could make to the sampling method.

6 The heights, in metres, of 20 ostriches are listed below:

1.8, 1.9, 2.3, 1.7, 2.1, 2.0, 2.5, 2.7, 2.5, 2.6, 2.3, 2.2, 2.4, 2.3, 2.2, 2.5, 1.9, 2.0, 2.2, 2.5

   a Take an opportunity sample of size five from the data.

   b Starting from the second data value, take a systematic sample of size five from the data.

   c Calculate the mean height for each sample.

   d State, with reasons, which sampling method is likely to be more reliable.

> **Hint** An example of an opportunity sample from this data would be to select the first five heights from the list.

1.4 Types of data

- **Variables or data associated with numerical observations are called quantitative variables or quantitative data.**

For example, you can give a number to shoe size so shoe size is a quantitative variable.

- **Variables or data associated with non-numerical observations are called qualitative variables or qualitative data.**

For example, you can't give a number to hair colour (blonde, red, brunette). Hair colour is a qualitative variable.

- **A variable that can take any value in a given range is a continuous variable.**

For example, time can take any value, e.g. 2 seconds, 2.1 seconds, 2.01 seconds etc.

- **A variable that can take only specific values in a given range is a discrete variable.**

For example, the number of girls in a family is a discrete variable as you can't have 2.65 girls in a family.

Large amounts of data can be displayed in a frequency table or as grouped data.

- **When data is presented in a grouped frequency table, the specific data values are not shown. The groups are more commonly known as classes.**
  - **Class boundaries tell you the maximum and minimum values that belong in each class.**
  - **The midpoint is the average of the class boundaries.**
  - **The class width is the difference between the upper and lower class boundaries.**

Example 4

The lengths, x mm, to the nearest mm, of the forewings of a random sample of male adult butterflies are measured and shown in the table.

Length of forewing (mm)	Number of butterflies, f
30–31	2
32–33	25
34–36	30
37–39	13

a State whether length is
   i quantitative or qualitative
   ii discrete or continuous.

b Write down the class boundaries, midpoint and class width for the class 34–36.

a i Quantitative
  ii Continuous

b Class boundaries 33.5 mm, 36.5 mm
  Midpoint = $\frac{1}{2}$(33.5 + 36.5) = 35 mm
  Class width = 36.5 − 33.5 = 3 mm

Watch out Be careful when finding class boundaries for continuous data. The data values have been rounded to the nearest mm, so the upper class boundary for the 30–31 mm class is 31.5 mm.

Exercise 1D

1 State whether each of the following variables is qualitative or quantitative.
   a Height of a tree
   b Colour of car
   c Time waiting in a queue
   d Shoe size
   e Names of pupils in a class

2 State whether each of the following quantitative variables is continuous or discrete.
   a Shoe size
   b Length of leaf
   c Number of people on a bus
   d Weight of sugar
   e Time required to run 100 m
   f Lifetime in hours of torch batteries

3 Explain why:
   a 'Type of tree' is a qualitative variable
   b 'The number of pupils in a class' is a discrete quantitative variable
   c 'The weight of a collie dog' is a continuous quantitative variable.

4 The distribution of the masses of two-month-old lambs is shown in the grouped frequency table.

Mass, m (kg)	Frequency
$1.2 \leqslant m < 1.3$	8
$1.3 \leqslant m < 1.4$	28
$1.4 \leqslant m < 1.5$	32
$1.5 \leqslant m < 1.6$	22

Hint The class boundaries are given using inequalities, so the values given in the table are the actual class boundaries.

   a Write down the class boundaries for the third group.
   b Work out the midpoint of the second group.
   c Work out the class width of the first group.

1.5 The large data set

You will need to answer questions based on real data in your exam. Some of these questions will be based on weather data from the **large data set** provided by Edexcel.

The data set consists of weather data samples provided by the Met Office for five UK weather stations and three overseas weather stations over two set periods of time: May to October 1987 and May to October 2015. The weather stations are labelled on the maps below.

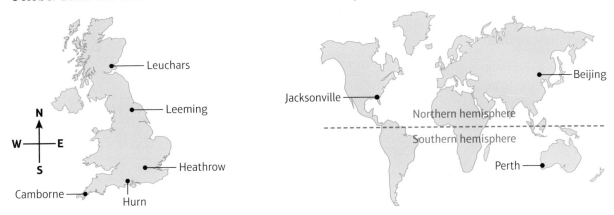

The large data set contains data for a number of different variables at each weather station:

- **Daily mean temperature** in °C – this is the average of the hourly temperature readings during a 24-hour period.
- **Daily total rainfall** including solid precipitation such as snow and hail, which is melted before being included in any measurements – amounts less than 0.05 mm are recorded as 'tr' or 'trace'
- **Daily total sunshine** recorded to the nearest tenth of an hour
- **Daily mean wind direction and windspeed** in knots, averaged over 24 hours from midnight to midnight. Mean wind directions are given as bearings and as cardinal (compass) directions. The data for mean windspeed is also categorised according to the **Beaufort scale**

Beaufort scale	Descriptive term	Average speed at 10 metres above ground
0	Calm	Less than 1 knot
1–3	Light	1 to 10 knots
4	Moderate	11 to 16 knots
5	Fresh	17 to 21 knots

Notation A **knot** (kn) is a 'nautical mile per hour'. 1 kn = 1.15 mph.

- **Daily maximum gust** in knots – this is the highest instantaneous windspeed recorded. The direction from which the maximum gust was blowing is also recorded
- **Daily maximum relative humidity**, given as a percentage of air saturation with water vapour. Relative humidities above 95% give rise to misty and foggy conditions

Watch out For the overseas locations, the only data recorded are:
- Daily mean temperature
- Daily total rainfall
- Daily mean pressure
- Daily mean windspeed

- **Daily mean cloud cover** measured in 'oktas' or eighths of the sky covered by cloud
- **Daily mean visibility** measured in decametres (Dm). This is the greatest horizontal distance at which an object can be seen in daylight
- **Daily mean pressure** measured in hectopascals (hPa)

Any missing data values are indicated in the large data set as n/a or 'not available'.

Data from Hurn for the first days of June 1987 is shown to the right.

You are expected to be able to take a sample from the large data set, identify different types of data and calculate statistics from the data.

- **If you need to do calculations on the large data set in your exam, the relevant extract from the data set will be provided.**

HURN					© Crown Copyright Met Office 1987	
Date	Daily mean temperature (°C)	Daily total rainfall (mm)	Daily total sunshine (hrs)	Daily mean windspeed (kn)	Daily mean windspeed (Beaufort conversion)	Daily maximum gust (kn)
01/6/1987	15.1	0.6	4.5	7	Light	19
02/6/1987	12.5	4.7	0	7	Light	22
03/6/1987	13.8	tr	5.6	11	Moderate	25
04/6/1987	15.5	5.3	7.8	7	Light	17
05/6/1987	13.1	19.0	0.5	10	Light	33
06/6/1987	13.8	0	8.9	19	Fresh	46
07/6/1987	13.2	tr	3.8	11	Moderate	27
08/6/1987	12.9	1	1.7	9	Light	19
09/6/1987	11.2	tr	5.4	6	Light	19
10/6/1987	9.2	1.3	9.7	4	Light	n/a
11/6/1987	12.6	0	12.5	6	Light	18
12/6/1987	10.4	0	11.9	5	Light	n/a
13/6/1987	9.6	0	8.6	5	Light	15
14/6/1987	10.2	0	13.1	5	Light	18
15/6/1987	9.2	3.7	7.1	4	Light	25
16/6/1987	10.4	5.6	8.3	6	Light	25
17/6/1987	12.8	0.1	5.3	10	Light	27
18/6/1987	13.0	7.4	3.2	9	Light	24
19/6/1987	14.0	tr	0.4	12	Moderate	33
20/6/1987	12.6	0	7.7	6	Light	17

Example 5

Look at the extract from the large data set given above.

a Describe the type of data represented by daily total rainfall.

Alison is investigating daily maximum gust. She wants to select a sample of size 5 from the first 20 days in Hurn in June 1987. She uses the first two digits of the date as a sampling frame and generates five random numbers between 1 and 20.

b State the type of sample selected by Alison.

c Explain why Alison's process might not generate a sample of size 5.

a Continuous quantitative data.
b Simple random sample
c Some of the data values are not available (n/a).

Watch out Although you won't need to recall specific data values from the large data set in your exam, you will need to know the limitations of the data set and the approximate range of values for each variable.

Example 6

Using the extract from the large data set on the previous page, calculate:

a the mean daily mean temperature for the first five days of June in Hurn in 1987

b the median daily total rainfall for the week of 14th June to 20th June inclusive.

The median daily total rainfall for the same week in Perth was 19.0 mm. Karl states that more southerly countries experience higher rainfall during June.

c State with a reason whether your answer to part **b** supports this statement.

a 15.1 + 12.5 + 13.8 + 15.5 + 13.1 = 70.0
70.0 ÷ 5 = 14.0 °C (1 d.p.)

The mean is the sum of the data values divided by the number of data values. The data values are given to 1 d.p. so give your answer to the same degree of accuracy.

b The values are: 0, 3.7, 5.6, 0.1, 7.4, tr, 0
In ascending order: 0, 0, tr, 0.1, 3.7, 5.6, 7.4
The median is the middle value so 0.1 mm.

Trace amounts are slightly larger than 0. If you need to do a numerical calculation involving a trace amount you can treat it as 0.

c Perth is in Australia, which is south of the UK, and the median rainfall was higher (19.0 mm > 0.1 mm). However, this is a very small sample from a single location in each country so does not provide enough evidence to support Karl's statement.

Online Use your calculator to find the mean and median of discrete data.

Problem-solving Don't just look at the numerical values. You also need to consider whether the sample is large enough, and whether there are other geographical factors which could affect rainfall in these two locations.

Exercise 1E

1 From the eight weather stations featured in the large data set, write down:
   a the station which is furthest north
   b the station which is furthest south
   c an inland station
   d a coastal station
   e an overseas station.

2 Explain, with reasons, whether daily maximum relative humidity is a discrete or continuous variable.

Questions 3 and 4 in this exercise use the following extracts from the large data set.

LEEMING
© Crown Copyright Met Office 2015

Date	Daily mean temperature (°C)	Daily total rainfall (mm)	Daily total sunshine (hrs)	Daily mean windspeed (kn)
01/06/2015	8.9	10	5.1	15
02/06/2015	10.7	tr	8.9	17
03/06/2015	12.0	0	10.0	8
04/06/2015	11.7	0	12.8	7
05/06/2015	15.0	0	8.9	9
06/06/2015	11.6	tr	5.4	17
07/06/2015	12.6	0	13.9	10
08/06/2015	9.4	0	9.7	7
09/06/2015	9.7	0	12.1	5
10/06/2015	11.0	0	14.6	4

HEATHROW
© Crown Copyright Met Office 2015

Date	Daily mean temperature (°C)	Daily total rainfall (mm)	Daily total sunshine (hrs)	Daily mean windspeed (kn)
01/06/2015	12.1	0.6	4.1	15
02/06/2015	15.4	tr	1.6	18
03/06/2015	15.8	0	9.1	9
04/06/2015	16.1	0.8	14.4	6
05/06/2015	19.6	tr	5.3	9
06/06/2015	14.5	0	12.3	12
07/06/2015	14.0	0	13.1	5
08/06/2015	14.0	tr	6.4	7
09/06/2015	11.4	0	2.5	10
10/06/2015	14.3	0	7.2	10

(P) **3 a** Work out the mean of the daily total sunshine for the first 10 days of June 2015 in:
   i Leeming
   ii Heathrow.

 b Work out the range of the daily total sunshine for the first 10 days of June 2015 in:
   i Leeming
   ii Heathrow.

 c Supraj says that the further north you are, the fewer the number of hours of sunshine. State, with reasons, whether your answers to parts **a** and **b** support this conclusion.

> **Hint** State in your answer whether Leeming is north or south of Heathrow.

(P) **4** Calculate the mean daily total rainfall in Heathrow for the first 10 days of June 2015. Explain clearly how you dealt with the data for 2/6/2015, 5/6/2015 and 8/6/2015.

(P) **5** Dominic is interested in seeing how the average monthly temperature changed over the summer months of 2015 in Jacksonville. He decides to take a sample of two days every month and average the temperatures before comparing them.

   a Give one reason why taking two days a month might be:

     i a good sample size

     ii a poor sample size.

   b He chooses the first day of each month and the last day of each month. Give a reason why this method of choosing days might not be representative.

   c Suggest a better way that he can choose his sample of days.

(P) **6** The table shows the mean daily temperatures at each of the eight weather stations for August 2015:

	Camborne	Heathrow	Hurn	Leeming	Leuchars	Beijing	Jacksonville	Perth
Mean daily mean temp (°C)	15.4	18.1	16.2	15.6	14.7	26.6	26.4	13.6

© Crown Copyright Met Office

   a Give a geographical reason why the temperature in August might be lower in Perth than in Jacksonville.

   b Comment on whether this data supports the conclusion that coastal locations experience lower average temperatures than inland locations.

(P) **7** Brian calculates the mean cloud coverage in Leeming in September 1987. He obtains the answer 9.3 oktas. Explain how you know that Brian's answer is incorrect.

(E/P) **8** The large data set provides data for 184 consecutive days in 1987. Marie is investigating daily mean windspeeds in Camborne in 1987.

   a Describe how Marie could take a systematic sample of 30 days from the data for Camborne in 1987. **(3 marks)**

   b Explain why Marie's sample would not necessarily give her 30 data points for her investigation. **(1 mark)**

Large data set

You will need access to the large data set and spreadsheet software to answer these questions.

1 a Find the mean daily mean pressure in Beijing in October 1987.

b Find the median daily rainfall in Jacksonville in July 2015.

c i Draw a grouped frequency table for the daily mean temperature in Heathrow in July and August 2015. Use intervals $10 \leqslant t < 15$, etc.

ii Draw a histogram to display this data.

iii Draw a frequency polygon for this data.

> **Hint** You can use the **CountIf** command in a spreadsheet to work out the frequency for each class.

2 a i Take a simple random sample of size 10 from the data for daily mean windspeed in Leeming in 1987.

ii Work out the mean of the daily windspeeds using your sample.

b i Take a sample of the last 10 values from the data for daily mean windspeed in Leuchars in 1987.

ii Work out the mean of the daily mean windspeeds using your sample.

c State, with reasons, which of your samples is likely to be more representative.

d Suggest two improvements to the sampling methods suggested in part **a**.

e Use an appropriate sampling method and sample size to estimate the mean windspeeds in Leeming and Leuchars in 1987. State with a reason whether your calculations support the statement 'Coastal locations are likely to have higher average windspeeds than inland locations'.

Mixed exercise 1

1 The table shows the daily mean temperature recorded on the first 15 days in May 1987 at Heathrow.

Day of month	1	2	3	4	5	6	7	8	9	10	11	12	13	14	15
Daily mean temp (°C)	14.6	8.8	7.2	7.3	10.1	11.9	12.2	12.1	15.2	11.1	10.6	12.7	8.9	10.0	9.5

© Crown Copyright Met Office

a Use an opportunity sample of the first 5 dates in the table to estimate the mean daily mean temperature at Heathrow for the first 15 days of May 1987.

b Describe how you could use the random number function on your calculator to select a simple random sample of 5 dates from this data.

> **Hint** Make sure you describe your sampling frame.

c Use a simple random sample of 5 dates to estimate the mean daily mean temperature at Heathrow for the first 15 days of May 1987.

d Use all 15 dates to calculate the mean daily mean temperature at Heathrow for the first 15 days of May 1987. Comment on the reliability of your two samples.

2 a Give one advantage and one disadvantage of using:

i a census   **ii** a sample survey.

b It is decided to take a sample of 100 from a population consisting of 500 elements. Explain how you would obtain a simple random sample from this population.

3 a Explain briefly what is meant by:
  i a population   **ii** a sampling frame.

 b A market research organisation wants to take a sample of:
  i owners of diesel motor cars in the UK
  ii persons living in Oxford who suffered injuries to the back during July 1996.

  Suggest a suitable sampling frame in each case.

4 Write down one advantage and one disadvantage of using:

 a stratified sampling   **b** simple random sampling.

5 The managing director of a factory wants to know what the workers think about the factory canteen facilities. 100 people work in the offices and 200 work on the shop floor.

 The factory manager decides to ask the people who work in the offices.

 a Suggest a reason why this is likely to produce a biased sample.

 b Explain briefly how the factory manager could select a sample of 30 workers using:
  i systematic sampling   **ii** stratified sampling   **iii** quota sampling.

6 There are 64 girls and 56 boys in a school.

 Explain briefly how you could take a random sample of 15 pupils using:

 a simple random sampling   **b** stratified sampling.

7 As part of her statistics project, Deepa decided to estimate the amount of time A-level students at her school spent on private study each week. She took a random sample of students from those studying arts subjects, science subjects and a mixture of arts and science subjects. Each student kept a record of the time they spent on private study during the third week of term.

 a Write down the name of the sampling method used by Deepa.

 b Give a reason for using this method and give one advantage this method has over simple random sampling.

8 A conservationist is collecting data on African springboks. She catches the first five springboks she finds and records their masses.

 a State the sampling method used.

 b Give one advantage of this type of sampling method.

 The data is given below:
 70 kg   76 kg   82 kg   74 kg   78 kg.

 c State, with a reason, whether this data is discrete or continuous.

 d Calculate the mean mass.

 A second conservationist collects data by selecting one springbok in each of five locations.
 The data collected is given below:

 79 kg   86 kg   90 kg   68 kg   75 kg.

 e Calculate the mean mass for this sample.

 f State, with a reason, which mean mass is likely to be a more reliable estimate of the mean mass of African springboks.

 g Give one improvement the second conservationist could make to the sampling method.

E **9** Data on the daily total rainfall in Beijing during 2015 is gathered from the large data set. The daily total rainfall (in mm) on the first of each month is listed below:

May 1st	9.0
June 1st	0.0
July 1st	1.0
August 1st	32.0
September 1st	4.1
October 1st	3.0

  a State, with a reason, whether or not this sample is random. **(1 mark)**

  b Suggest two alternative sampling methods and give one advantage and one disadvantage of each in this context. **(2 marks)**

  c State, with a reason, whether the data is discrete or continuous. **(1 mark)**

  d Calculate the mean of the six data values given above. **(1 mark)**

  e Comment on the reliability of this value as an estimate for the mean daily total rainfall in Beijing during 2015. **(1 mark)**

Large data set

You will need access to the large data set and spreadsheet software to answer these questions.

a Take a systematic sample of size 18 for the daily maximum relative humidity in Camborne during 1987.

b Give one advantage of using a systematic sample in this context.

c Use your sample to find an estimate for the mean daily maximum relative humidity in Camborne during 1987.

d Comment on the reliability of this estimate. Suggest one way in which the reliability can be improved.

Summary of key points

1 · In statistics, a **population** is the whole set of items that are of interest.
 · A **census** observes or measures every member of a population.

2 · A sample is a selection of observations taken from a subset of the population which is used to find out information about the population as a whole.
 · Individual units of a population are known as **sampling units**.
 · Often sampling units of a population are individually named or numbered to form a list called a **sampling frame**.

3 · A **simple random sample** of size n is one where every sample of size n has an equal chance of being selected.
 · In **systematic sampling**, the required elements are chosen at regular intervals from an ordered list.
 · In **stratified sampling**, the population is divided into mutually exclusive strata (males and females, for example) and a random sample is taken from each.
 · In **quota sampling**, an interviewer or researcher selects a sample that reflects the characteristics of the whole population.
 · **Opportunity sampling** consists of taking the sample from people who are available at the time the study is carried out and who fit the criteria you are looking for.

4 · Variables or data associated with numerical observations are called **quantitative variables** or **quantitative data**.
 · Variables or data associated with non-numerical observations are called **qualitative variables** or **qualitative data**.

5 · A variable that can take any value in a given range is a **continuous variable**.
 · A variable that can take only specific values in a given range is a **discrete variable**.

6 · When data is presented in a grouped frequency table, the specific data values are not shown. The groups are more commonly known as **classes**.
 · Class boundaries tell you the maximum and minimum values that belong in each class.
 · The midpoint is the average of the class boundaries.
 · The class width is the difference between the upper and lower class boundaries.

7 If you need to do calculations on the large data set in your exam, the relevant extract from the data set will be provided.

2 Measures of location and spread

Objectives

After completing this chapter you should be able to:

● Calculate measures of central tendency such as the mean, median and mode
→ pages 21–25

● Calculate measures of location such as percentiles and deciles
→ pages 25–28

● Calculate measures of spread such as range, interquartile range and interpercentile range
→ pages 28–29

● Calculate variance and standard deviation
→ pages 30–33

● Understand and use coding
→ pages 33–36

Prior knowledge check

1 State whether each of these variables is qualitative or quantitative:
 a Colour of car
 b Miles travelled by a cyclist
 c Favourite type of pet
 d Number of brothers and sisters.
 ← Section 1.4

2 State whether each of these variables is discrete or continuous:
 a Number of pets owned
 b Distance walked by ramblers
 c Fuel consumption of lorries
 d Number of peas in a pod
 e Times taken by a group of athletes to run 1500 m.
 ← Section 1.4

3 Find the mean, median, mode and range of the data shown in this frequency table.

Number of peas in a pod	3	4	5	6	7
Frequency	4	7	11	18	6

← GCSE Mathematics

Wildlife biologists use statistics such as mean wingspan and standard deviation to compare populations of endangered birds in different habitats.
→ Mixed exercise Q12

2.1 Measures of central tendency

A **measure of location** is a single value which describes a position in a data set. If the single value describes the centre of the data, it is called a **measure of central tendency**. You should already know how to work out the **mean**, **median** and **mode** of a set of ungrouped data and from ungrouped frequency tables.

- **The mode or modal class is the value or class that occurs most often.**

- **The median is the middle value when the data values are put in order.**

- **The mean can be calculated using the formula $\bar{x} = \dfrac{\Sigma x}{n}$.**

Notation
- $\bar{x}$ represents the **mean** of the data. You say 'x bar'.
- Σx represents the sum of the data values.
- n is the number of data values.

Example 1

The mean of a sample of 25 observations is 6.4. The mean of a second sample of 30 observations is 7.2. Calculate the mean of all 55 observations.

For the first set of observations:

$\bar{x} = \dfrac{\Sigma x}{n}$ so $6.4 = \dfrac{\Sigma x}{25}$

$\Sigma x = 6.4 \times 25 = 160$ Sum of data values = mean × number of data values.

For the second set of observations:

$\bar{y} = \dfrac{\Sigma y}{m}$ so $7.2 = \dfrac{\Sigma y}{30}$

$\Sigma y = 7.2 \times 30 = 216$

Mean $= \dfrac{160 + 216}{25 + 30} = 6.84$ (2 d.p.)

Notation You can use x and y to represent two different data sets. You need to use different letters for the number of observations in each data set.

You need to decide on the best measure to use in particular situations.

- Mode   This is used when data is qualitative, or quantitative with either a single mode or two modes (bimodal). It is not very informative if each value occurs only once.

- Median   This is used for quantitative data. It is usually used when there are extreme values, as they do not affect it.

- Mean   This is used for quantitative data and uses all the pieces of data. It therefore gives a true measure of the data. However, it is affected by extreme values.

You can calculate the mean, median and mode for discrete data presented in a frequency table.

- **For data given in a frequency table, the mean can be calculated using the formula $\bar{x} = \dfrac{\Sigma xf}{\Sigma f}$.**

Notation
- Σxf is the sum of the products of the data values and their frequencies.
- Σf is the sum of the frequencies.

Example 2

Rebecca records the shirt collar size, x, of the male students in her year. The results are shown in the table.

Shirt collar size	15	15.5	16	16.5	17
Number of students	3	17	29	34	12

Find for this data:

a the mode   **b** the median   **c** the mean.

d Explain why a shirt manufacturer might use the mode when planning production numbers.

a Mode = 16.5

16.5 is the collar size with the highest frequency.

b There are 95 observations

so the median is the $\frac{95 + 1}{2} = 48$th.

There are 20 observations up to 15.5 and 49 observations up to 16.

Median = 16

The 48th observation is therefore 16.

c $\bar{x} = \dfrac{15 \times 3 + 15.5 \times 17 + 16 \times 29 + 16.5 \times 34 + 17 \times 12}{95}$

$= \dfrac{45 + 263.5 + 464 + 561 + 204}{95} = \dfrac{1537.5}{95} = 16.2$

Online You can input a frequency table into your calculator, and calculate the mean and median without having to enter the whole calculation.

d The mode is an actual data value and gives the manufacturer information on the most common size worn/purchased.

The mean is not one of the data values and the median is not necessarily indicative of the most popular collar size.

Exercise 2A

1 Meryl collected wild mushrooms every day for a week. When she got home each day she weighed them to the nearest 100 g. The weights are shown below:

     500   700   400   300   900   700   700

   a Write down the mode for this data.

   b Calculate the mean for this data.

   c Find the median for this data.

On the next day, Meryl collects 650 g of wild mushrooms.

   d Write down the effect this will have on the mean, the mode and the median.

Hint Try to answer part **d** without recalculating the averages. You could recalculate to check your answer.

2 Joe collects six pieces of data, x_1, x_2, x_3, x_4, x_5 and x_6. He works out that Σx is 256.2.

   a Calculate the mean for this data.

He collects another piece of data. It is 52.

   b Write down the effect this piece of data will have on the mean.

3 From the large data set, the daily mean visibility, v metres, for Leeming in May and June 2015 was recorded each day. The data is summarised as follows:

May: $n = 31$, $\Sigma v = 724\,000$

June: $n = 30$, $\Sigma v = 632\,000$

a Calculate the mean visibility in each month.

b Calculate the mean visibility for the total recording period.

> **Hint** You don't need to refer to the actual large data set. All the data you need is given with the question.

4 A small workshop records how long it takes, in minutes, for each of their workers to make a certain item. The times are shown in the table.

Worker	A	B	C	D	E	F	G	H	I	J
Time in minutes	7	12	10	8	6	8	5	26	11	9

a Write down the mode for this data.

b Calculate the mean for this data.

c Find the median for this data.

d The manager wants to give the workers an idea of the average time they took. Write down, with a reason, which of the answers to **a**, **b** and **c** she should use.

5 The frequency table shows the number of breakdowns, b, per month recorded by a road haulage firm over a certain period of time.

Breakdowns	0	1	2	3	4	5
Frequency	8	11	12	3	1	1

a Write down the modal number of breakdowns.

b Find the median number of breakdowns.

c Calculate the mean number of breakdowns.

d In a brochure about how many loads reach their destination on time, the firm quotes one of the answers to **a**, **b** or **c** as the number of breakdowns per month for its vehicles. Write down which of the three answers the firm should quote in the brochure.

6 The table shows the frequency distribution for the number of petals in the flowers of a group of celandines.

Number of petals	5	6	7	8	9
Frequency	8	57	29	3	1

Calculate the mean number of petals.

(P) **7** A naturalist is investigating how many eggs the endangered kakapo bird lays in each brood cycle. The results are given in this frequency table.

Number of eggs	1	2	3
Frequency	7	p	2

If the mean number of eggs is 1.5, find the value of p.

> **Problem-solving**
> Use the formula for the mean of an ungrouped frequency table to write an equation involving p.

You can calculate the mean, the class containing the median and the modal class for continuous data presented in a grouped frequency table by finding the midpoint of each class interval.

Example 3

The length x mm, to the nearest mm, of a random sample of pine cones is measured. The data is shown in the table.

Length of pine cone (mm)	30–31	32–33	34–36	37–39
Frequency	2	25	30	13

a Write down the modal class. **b** Estimate the mean. **c** Find the median class.

a Modal class = 34–36

b Mean = $\dfrac{30.5 \times 2 + 32.5 \times 25 + 35 \times 30 + 38 \times 13}{70}$

$= 34.54$

c There are 70 observations so the median is the 35.5th. The 35.5th observation will lie in the class 34–36.

The modal class is the class with the highest frequency.

Use $\bar{x} = \dfrac{\Sigma xf}{\Sigma f}$, taking the midpoint of each class interval as the value of x. The answer is an estimate because you don't know the exact data values.

← **Section 1.4**

Exercise 2B

1 The weekly wages (to the nearest £) of the production line workers in a small factory is shown in the table.

a Write down the modal class.

b Calculate an estimate of the mean wage.

c Write down the interval containing the median.

Weekly wage (£)	Frequency
175–225	4
226–300	8
301–350	18
351–400	28
401–500	7

E **2** The noise levels at 30 locations near an outdoor concert venue were measured to the nearest decibel. The data collected is shown in the grouped frequency table.

Noise (decibels)	65–69	70–74	75–79	80–84	85–89	90–94	95–99
Frequency	1	4	6	6	8	4	1

a Calculate an estimate of the mean noise level. **(1 mark)**

b Explain why your answer to part **a** is an estimate. **(1 mark)**

E **3** The table shows the daily mean temperature at Heathrow in October 1987 from the large data set.

Temp (°C)	$6 \leqslant t < 8$	$8 \leqslant t < 10$	$10 \leqslant t < 12$	$12 \leqslant t < 14$	$14 \leqslant t < 16$	$16 \leqslant t < 18$
Frequency	3	7	9	7	3	2

© Crown Copyright Met Office

a Write down the modal class. **(1 mark)**

b Calculate an estimate for the mean daily mean temperature. **(1 mark)**

(P) **4** Two DIY shops (A and B) recorded the ages of their workers.

Age of worker	16–25	26–35	36–45	46–55	56–65	66–75
Frequency A	5	16	14	22	26	14
Frequency B	4	12	10	28	25	13

By comparing estimated means for each shop, determine which shop is better at employing older workers.

Problem-solving

Since age is always rounded **down**, the class boundaries for the 16–25 group are 16 and 26. This means that the midpoint of the class is 21.

2.2 Other measures of location

The median describes the middle of the data set. It splits the data set into two equal (50%) halves.

You can calculate other measures of location such as **quartiles** and **percentiles**.

The **lower quartile** is one-quarter of the way through the data set.

This is the median value.

The **upper quartile** is three-quarters of the way through the data set.

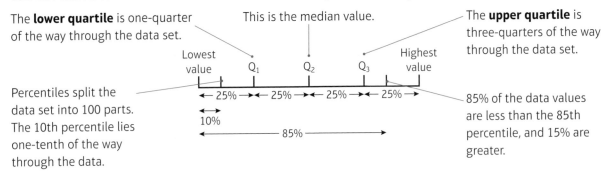

Percentiles split the data set into 100 parts. The 10th percentile lies one-tenth of the way through the data.

85% of the data values are less than the 85th percentile, and 15% are greater.

Use these rules to find the upper and lower quartiles for **discrete data**.

■ **To find the lower quartile for discrete data, divide *n* by 4. If this is a whole number, the lower quartile is halfway between this data point and the one above. If it is not a whole number, round *up* and pick this data point.**

■ **To find the upper quartile for discrete data, find $\frac{3}{4}$ of *n*. If this is a whole number, the upper quartile is halfway between this data point and the one above. If it is not a whole number, round *up* and pick this data point.**

Notation Q_1 is the lower quartile, Q_2 is the median and Q_3 is the upper quartile.

Example 4

From the large data set, the daily maximum gust (knots) during the first 20 days of June 2015 is recorded in Hurn. The data is shown below:

14	15	17	17	18	18	19	19	22	22
23	23	23	24	25	26	27	28	36	39

© Crown Copyright Met Office

Find the median and quartiles for this data.

$$Q_2 = \frac{20 + 1}{2}\text{th value} = 10.5\text{th value}$$

$$Q_2 = \frac{22 + 23}{2} = 22.5 \text{ knots}$$

$$Q_1 = 5.5\text{th value}$$

$$Q_1 = 18 \text{ knots}$$

$$Q_3 = 15.5\text{th value}$$

$$Q_3 = 25.5 \text{ knots}$$

Q_2 is the median. It lies halfway between the 10th and 11th data values (which are 22 knots and 23 knots respectively).

$\frac{20}{4} = 5$ so the lower quartile is halfway between the 5th and 6th data values.

$\frac{3 \times 20}{4} = 15$ so the upper quartile is halfway between the 15th and 16th data values.

When data are presented in a grouped frequency table you can use a technique called **interpolation** to estimate the median, quartiles and percentiles. When you use interpolation, you are assuming that the data values are **evenly distributed** within each class.

Watch out For **grouped continuous** data, or data presented in a cumulative frequency table:

$$Q_1 = \frac{n}{4}\text{th data value}$$

$$Q_2 = \frac{n}{2}\text{th data value}$$

$$Q_3 = \frac{3n}{4}\text{th data value}$$

Example 5

The length of time (to the nearest minute) spent on the internet each evening by a group of students is shown in the table.

Length of time spent on internet (minutes)	30–31	32–33	34–36	37–39
Frequency	2	25	30	13

a Find an estimate for the upper quartile.

b Find an estimate for the 10th percentile.

a Upper quartile: $\frac{3 \times 70}{4} = 52.5\text{th value}$

Using interpolation:

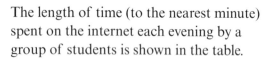

$$\frac{Q_3 - 33.5}{36.5 - 33.5} = \frac{52.5 - 27}{57 - 27}$$

$$\frac{Q_3 - 33.5}{3} = \frac{25.5}{30}$$

$$Q_3 = 36.05$$

b The 10th percentile is the 7th data value.

$$\frac{P_{10} - 31.5}{33.5 - 31.5} = \frac{7 - 2}{27 - 2}$$

$$\frac{P_{10} - 31.5}{2} = \frac{5}{25}$$

$$P_{10} = 31.9$$

The endpoints on the line represent the class boundaries.

The values on the bottom are the cumulative frequencies for the previous classes and this class.

Problem-solving

Use proportion to estimate Q_3. The 52.5th value lies $\frac{52.5 - 27}{57 - 27}$ of the way into the class, so Q_3 lies $\frac{Q_3 - 33.5}{36.5 - 33.5}$ of the way between the class boundaries. Equate these two fractions to form an equation and solve to find Q_3.

Notation You can write the 10th percentile as P_{10}.

Exercise 2C

1 From the large data set, the daily mean pressure (hPa) during the last 16 days of July 2015 in Perth is recorded. The data is given below:

1024	1022	1021	1013	1009	1018	1017	1024
1027	1029	1031	1025	1017	1019	1017	1014

  a Find the median pressure for that period.

  b Find the lower and upper quartiles.

2 Rachel records the number of CDs in the collections of students in her year. The results are in the table below.

Number of CDs	35	36	37	38	39
Frequency	3	17	29	34	12

  Find Q_1, Q_2 and Q_3.

Hint This an ungrouped frequency table so you do not need to use interpolation. Use the rules for finding the median and quartiles of **discrete** data.

(E) **3** A hotel is worried about the reliability of its lift. It keeps a weekly record of the number of times it breaks down over a period of 26 weeks. The data collected is summarised in the table opposite.

Use interpolation to estimate the median number of breakdowns. **(2 marks)**

Number of breakdowns	Frequency
0–1	18
2–3	7
4–5	1

4 The weights of 31 Jersey cows were recorded to the nearest kilogram. The weights are shown in the table.

  a Find an estimate for the median weight.

  b Find the lower quartile, Q_1.

  c Find the upper quartile, Q_3.

Weight of cattle (kg)	300–349	350–399	400–449	450–499	500–549
Frequency	3	6	10	7	5

  d Interpret the meaning of the value you have found for the upper quartile in part **c**.

(E) **5** A roadside assistance firm kept a record over a week of the amount of time, in minutes, people were kept waiting for assistance. The times are shown below.

Time waiting, t (minutes)	$20 \leqslant t < 30$	$30 \leqslant t < 40$	$40 \leqslant t < 50$	$50 \leqslant t < 60$	$60 \leqslant t < 70$
Frequency	6	10	18	13	2

  a Find an estimate for the mean wait time. **(1 mark)**

  b Calculate the 65th percentile. **(2 marks)**

The firm writes the following statement for an advertisement:

> Only 10% of our customers have to wait longer than 56 minutes.

  c By calculating a suitable percentile, comment on the validity of this claim. **(3 marks)**

(E) **6** The table shows the recorded wingspans, in metres, of 100 endangered Californian condors.

Wingspan, w (m)	$1.0 \leqslant w < 1.5$	$1.5 \leqslant w < 2.0$	$2.0 \leqslant w < 2.5$	$2.5 \leqslant w < 3.0$	$3.0 \leqslant w$
Frequency	4	20	37	28	11

   a Estimate the 80th percentile and interpret the value.   **(3 marks)**

   b State why it is not possible to estimate the 90th percentile.   **(1 mark)**

2.3 Measures of spread

A measure of spread is a measure of how spread out the data is. Here are two simple measures of spread.

Notation   Measures of spread are sometimes called **measures of dispersion** or **measures of variation**.

- **The range is the difference between the largest and smallest values in the data set.**

- **The interquartile range (IQR) is the difference between the upper quartile and the lower quartile, $Q_3 - Q_1$.**

The range takes into account all of the data but can be affected by extreme values. The interquartile range is not affected by extreme values but only considers the spread of the middle 50% of the data.

- **The interpercentile range is the difference between the values for two given percentiles.**

The 10th to 90th interpercentile range is often used since it is not affected by extreme values but still considers 80% of the data in its calculation.

Example 6

The table shows the masses, in tonnes, of 120 African bush elephants.

Mass, m (t)	$4.0 \leqslant m < 4.5$	$4.5 \leqslant m < 5.0$	$5.0 \leqslant m < 5.5$	$5.5 \leqslant m < 6.0$	$6.0 \leqslant m < 6.5$
Frequency	13	23	31	34	19

Find estimates for:

a the range   **b** the interquartile range   **c** the 10th to 90th interpercentile range.

a Range is 6.5 − 4.0 = 2.5 tonnes.

The largest possible value is 6.5 and the smallest possible value is 4.0.

b Q_1 = 30th data value: 4.87 tonnes.
Q_3 = 90th data value: 5.84 tonnes.
The interquartile range is therefore
5.84 − 4.87 = 0.97 tonnes.

Use interpolation: $\dfrac{Q_1 - 4.5}{5.0 - 4.5} = \dfrac{30 - 13}{23}$

Use interpolation: $\dfrac{Q_3 - 5.5}{6.0 - 5.5} = \dfrac{90 - 67}{34}$

c 10th percentile = 12th data value:
4.46 tonnes.
90th percentile = 108th data value:
6.18 tonnes.
The 10th to 90th interpercentile range is
therefore 6.18 − 4.46 = 1.72 tonnes.

Use interpolation to find the 10th and 90th percentiles, then work out the difference between them.

Exercise 2D

(P) 1 The lengths of a number of slow worms were measured, to the nearest mm.
The results are shown in the table.

Lengths of slow worms (mm)	Frequency
125–139	4
140–154	4
155–169	2
170–184	7
185–199	20
200–214	24
215–229	10

a Work out how many slow worms were measured.

b Estimate the interquartile range for the lengths of the slow worms.

c Calculate an estimate for the mean length of slow worms.

d Estimate the number of slow worms whose length is more than one interquartile range above the mean.

Problem-solving

For part **d**, work out $\bar{x}$ + IQR, and determine which class interval it falls in. Then use proportion to work out how many slow worms from that class interval you need to include in your estimate.

(E) 2 The table shows the monthly income for workers in a factory.

Monthly income, x (£)	$900 \leqslant x < 1000$	$1000 \leqslant x < 1100$	$1100 \leqslant x < 1200$	$1200 \leqslant x < 1300$
Frequency	3	24	28	15

a Calculate the 34% to 66% interpercentile range. **(3 marks)**

b Estimate the number of data values that fall within this range. **(2 marks)**

(E) 3 A train travelled from Lancaster to Preston. The times, to the nearest minute, it took for the journey were recorded over a certain period. The times are shown in the table.

Time for journey (minutes)	15–16	17–18	19–20	21–22
Frequency	5	10	35	10

a Calculate the 5% to 95% interpercentile range. **(3 marks)**

b Estimate the number of data values that fall within this range. **(1 mark)**

(E/P) 4 From the large data set, the daily mean temperature (°C) for Leeming during the first 10 days of June 1987 is given below:

14.3   12.7   12.4   10.9   9.4   13.2   12.1   10.3   10.3   10.6

a Calculate the median and interquartile range. **(2 marks)**

The median daily mean temperature for Leeming during the first 10 days of May 1987 was 9.9 °C and the interquartile range was 3.9 °C.

b Compare the data for May with the data for June. **(2 marks)**

The 10% to 90% interpercentile range for the daily mean temperature for Leeming during July 1987 was 5.4 °C.

c Estimate the number of days in July 1987 on which the daily mean temperature fell within this range. **(1 mark)**

2.4 Variance and standard deviation

Another measure that can be used to work out the spread of a data set is the **variance**. This makes use of the fact that each data point deviates from the mean by the amount $x - \bar{x}$.

- **Variance** $= \dfrac{\Sigma(x - \bar{x})^2}{n} = \dfrac{\Sigma x^2}{n} - \left(\dfrac{\Sigma x}{n}\right)^2 = \dfrac{S_{xx}}{n}$

  where $S_{xx} = \Sigma(x - \bar{x})^2 = \Sigma x^2 - \dfrac{(\Sigma x)^2}{n}$

Notation S_{xx} is a **summary statistic**, which is used to make formulae easier to use and learn.

The second version of the formula, $\dfrac{\Sigma x^2}{n} - \left(\dfrac{\Sigma x}{n}\right)^2$, is easier to work with when given raw data. It can be thought of as 'the mean of the squares minus the square of the mean'.

The third version, $\dfrac{S_{xx}}{n}$, is easier to use if you can use your calculator to find S_{xx} quickly.

The units of the variance are the units of the data squared. You can find a related measure of spread that has the same units as the data.

- **The standard deviation is the square root of the variance:**

  $\sigma = \sqrt{\dfrac{\Sigma(x - \bar{x})^2}{n}} = \sqrt{\dfrac{\Sigma x^2}{n} - \left(\dfrac{\Sigma x}{n}\right)^2} = \sqrt{\dfrac{S_{xx}}{n}}$

Notation σ is the symbol we use for the standard deviation of a data set. Hence σ^2 is used for the variance.

Example 7

The marks gained in a test by seven randomly selected students are:

    3   4   6   2   8   8   5

Find the variance and standard deviation of the marks of the seven students.

$\Sigma x = 3 + 4 + 6 + 2 + 8 + 8 + 5 = 36$

$\Sigma x^2 = 9 + 16 + 36 + 4 + 64 + 64 + 25 = 218$

variance, $\sigma^2 = \dfrac{218}{7} - \left(\dfrac{36}{7}\right)^2 = 4.69$

standard deviation, $\sigma = \sqrt{4.69} = 2.17$

Use the 'mean of the squares minus the square of the mean'.

$\sigma^2 = \dfrac{\Sigma x^2}{n} - \left(\dfrac{\Sigma x}{n}\right)^2$

- **You can use these versions of the formulae for variance and standard deviation for grouped data that is presented in a frequency table:**

  - $\sigma^2 = \dfrac{\Sigma f(x - \bar{x})^2}{\Sigma f} = \dfrac{\Sigma fx^2}{\Sigma f} - \left(\dfrac{\Sigma fx}{\Sigma f}\right)^2$

  - $\sigma = \sqrt{\dfrac{\Sigma f(x - \bar{x})^2}{\Sigma f}} = \sqrt{\dfrac{\Sigma fx^2}{\Sigma f} - \left(\dfrac{\Sigma fx}{\Sigma f}\right)^2}$

    where f is the frequency for each group and Σf is the total frequency.

Example 8

Shamsa records the time spent out of school during the lunch hour to the nearest minute, x, of the female students in her year.
The results are shown in the table.

Time spent out of school (min)	35	36	37	38
Frequency	3	17	29	34

Calculate the standard deviation of the time spent out of school.

$\Sigma fx^2 = 3 \times 35^2 + 17 \times 36^2 + 29 \times 37^2$
$\qquad + 34 \times 38^2 = 114\,504$

$\Sigma fx = 3 \times 35 + 17 \times 36 + 29 \times 37$
$\qquad + 34 \times 38 = 3082$

$\Sigma f = 3 + 17 + 29 + 34 = 83$

$\sigma^2 = \dfrac{114\,504}{83} - \left(\dfrac{3082}{83}\right)^2 = 0.741\,47\ldots$

$\sigma = \sqrt{0.741\,47\ldots} = 0.861$ (3 s.f.)

> The values of Σfx^2, Σfx and Σf might be given with the question.

> σ^2 is the variance, and σ is the standard deviation.
> Use $\sigma^2 = \dfrac{\Sigma fx^2}{\Sigma f} - \left(\dfrac{\Sigma fx}{\Sigma f}\right)^2$

If the data is given in a grouped frequency table, you can calculate **estimates** for the variance and standard deviation of the data using the **midpoint** of each class interval.

Example 9

Andy recorded the length, in minutes, of each telephone call he made for a month. The data is summarised in the table below.

Length of telephone call (l min)	$0 < l \leq 5$	$5 < l \leq 10$	$10 < l \leq 15$	$15 < l \leq 20$	$20 < l \leq 60$	$60 < l \leq 70$
Frequency	4	15	5	2	0	1

Calculate an estimate of the standard deviation of the length of telephone calls.

Length of telephone call (l min)	Frequency	Midpoint x	fx	fx^2
$0 < l \leq 5$	4	2.5	$4 \times 2.5 = 10$	$4 \times 6.25 = 25$
$5 < l \leq 10$	15	7.5	112.5	843.75
$10 < l \leq 15$	5	12.5	62.5	781.25
$15 < l \leq 20$	2	17.5	35	612.5
$20 < l \leq 60$	0	40	0	0
$60 < l \leq 70$	1	65	65	4225
total	27		285	6487.5

$\Sigma fx^2 = 6487.5 \qquad \Sigma fx = 285 \qquad \Sigma f = 27$

$\sigma^2 = \dfrac{6487.5}{27} - \left(\dfrac{285}{27}\right)^2 = 128.858\,02$

$\sigma = \sqrt{128.858\,02} = 11.35$

> You can use a table like this to keep track of your working.

> **Online** Work this out in one go on your calculator. You might need to enter the values manually for the midpoint of each class interval.

Exercise 2E

1 Given that for a variable x: $\Sigma x = 24$ $\Sigma x^2 = 78$ $n = 8$
  Find:
  a the mean **b** the variance σ^2 **c** the standard deviation σ.

E **2** Ten collie dogs are weighed (w kg). The summary data for the weights is:
  $$\Sigma w = 241 \qquad \Sigma w^2 = 5905$$
  Use this summary data to find the standard deviation of the collies' weights. **(2 marks)**

3 Eight students' heights (h cm) are measured. They are as follows:
  165   170   190   180   175   185   176   184
  a Work out the mean height of the students.
  b Given $\Sigma h^2 = 254\,307$ work out the variance. Show all your working.
  c Work out the standard deviation.

P **4** For a set of 10 numbers: $\Sigma x = 50$ $\Sigma x^2 = 310$
  For a different set of 15 numbers: $\Sigma x = 86$ $\Sigma x^2 = 568$
  Find the mean and the standard deviation of the combined set of 25 numbers.

E **5** Nahab asks the students in his year group how much pocket money they get per week. The results, rounded to the nearest pound, are shown in the table.

Number of £s	8	9	10	11	12
Frequency	14	8	28	15	20

  a Use your calculator to work out the mean and standard deviation of the pocket money. Give units with your answer. **(3 marks)**
  b How many students received an amount of pocket money more than one standard deviation above the mean? **(2 marks)**

E **6** In a student group, a record was kept of the number of days of absence each student had over one particular term. The results are shown in the table.

Number of days absent	0	1	2	3	4
Frequency	12	20	10	7	5

  Use your calculator to work out the standard deviation of the number of days absent. **(2 marks)**

E/P **7** A certain type of machine contained a part that tended to wear out after different amounts of time. The time it took for 50 of the parts to wear out was recorded. The results are shown in the table.

Lifetime, h (hours)	$5 < h \leqslant 10$	$10 < h \leqslant 15$	$15 < h \leqslant 20$	$20 < h \leqslant 25$	$25 < h \leqslant 30$
Frequency	5	14	23	6	2

The manufacturer makes the following claim:

90% of the parts tested lasted longer than one standard deviation below the mean.

Comment on the accuracy of the manufacturer's claim, giving relevant numerical evidence.

Problem-solving

You need to calculate estimates for the mean and the standard deviation, then estimate the number of parts that lasted longer than one standard deviation below the mean.

(5 marks)

(E) **8** The daily mean windspeed, x (kn) for Leeming is recorded in June 2015. The summary data is:

$$\Sigma x = 243 \qquad \Sigma x^2 = 2317$$

a Use your calculator to work out the mean and the standard deviation of the daily mean windspeed in June 2015. **(2 marks)**

The highest recorded windspeed was 17 kn and the lowest recorded windspeed was 4 kn.

b Estimate the number of days in which the windspeed was greater than one standard deviation above the mean. **(2 marks)**

c State one assumption you have made in producing this estimate. **(1 mark)**

2.5 Coding

Coding is a way of simplifying statistical calculations. Each data value is coded to make a new set of data values which are easier to work with.

In your exam, you will usually have to code values using a formula like this: $y = \dfrac{x - a}{b}$

where a and b are constants that you have to choose or are given with the question.

When data is coded, different statistics change in different ways.

■ **If data is coded using the formula** $y = \dfrac{x - a}{b}$

 • **the mean of the coded data is given by** $\bar{y} = \dfrac{\bar{x} - a}{b}$

 • **the standard deviation of the coded data is given by** $\sigma_y = \dfrac{\sigma_x}{b}$**, where** σ_x **is the standard deviation of the original data.**

Hint You usually need to find the mean and standard deviation of the **original data** given the statistics for the **coded data**. You can rearrange the formulae as:

 • $\bar{x} = b\bar{y} + a$

 • $\sigma_x = b\sigma_y$

Example 10

A scientist measures the temperature, $x\,^\circ C$, at five different points in a nuclear reactor. Her results are given below:

$$332\,^\circ C \qquad 355\,^\circ C \qquad 306\,^\circ C \qquad 317\,^\circ C \qquad 340\,^\circ C$$

a Use the coding $y = \dfrac{x - 300}{10}$ to code this data.

b Calculate the mean and standard deviation of the coded data.

c Use your answer to part **b** to calculate the mean and standard deviation of the original data.

a

Original data, x	332	355	306	317	340
Coded data, y	3.2	5.5	0.6	1.7	4.0

When $x = 332$, $y = \dfrac{332 - 300}{10} = 3.2$.

b $\Sigma y = 15$, $\Sigma y^2 = 59.74$

$$\bar{y} = \frac{15}{5} = 3$$

$$\sigma_y^2 = \frac{59.74}{5} - \left(\frac{15}{5}\right)^2 = 2.948$$

$$\sigma_y = \sqrt{2.948} = 1.72 \text{ (3 s.f.)}$$

Substitute into $\bar{y} = \dfrac{\bar{x} - a}{b}$ and solve to find $\bar{x}$.

You could also use $\bar{x} = b\bar{y} + a$ with $a = 300$, $b = 10$ and $\bar{y} = 3$.

c $3 = \dfrac{\bar{x} - 300}{10}$ so $\bar{x} = 30 + 300 = 330\,°\text{C}$

$$1.72 = \frac{\sigma_x}{10} \text{ so } \sigma_x = 17.2\,°\text{C (3 s.f.)}$$

Substitute into $\sigma_y = \dfrac{\sigma_x}{b}$ and solve to find σ_x.

You could also use $\sigma_x = b\sigma_y$ with $\sigma_y = 1.72$ and $b = 10$.

Example 11

From the large data set, data on the maximum gust, g knots, is recorded in Leuchars during May and June 2015.

The data was coded using $h = \dfrac{g - 5}{10}$ and the following statistics found:

$$S_{hh} = 43.58 \qquad \bar{h} = 2 \qquad n = 61$$

Calculate the mean and standard deviation of the maximum gust in knots.

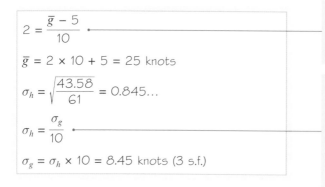

$$2 = \frac{\bar{g} - 5}{10}$$

$$\bar{g} = 2 \times 10 + 5 = 25 \text{ knots}$$

$$\sigma_h = \sqrt{\frac{43.58}{61}} = 0.845\ldots$$

$$\sigma_h = \frac{\sigma_g}{10}$$

$$\sigma_g = \sigma_h \times 10 = 8.45 \text{ knots (3 s.f.)}$$

Use the formula for the mean of a coded variable:

$\bar{h} = \dfrac{\bar{g} - a}{b}$ with $a = 5$ and $b = 10$.

Calculate the standard deviation of the coded data using $\sigma_h = \sqrt{\dfrac{S_{hh}}{n}}$, then use the formula for the standard deviation of a coded variable:

$\sigma_h = \dfrac{\sigma_g}{b}$ with $b = 10$.

Exercise 2F

1 A set of data values, x, is shown below:

$$110 \qquad 90 \qquad 50 \qquad 80 \qquad 30 \qquad 70 \qquad 60$$

a Code the data using the coding $y = \dfrac{x}{10}$.

b Calculate the mean of the coded data values.

c Use your answer to part **b** to calculate the mean of the original data.

2 A set of data values, x, is shown below:

    52   73   31   73   38   80   17   24

a Code the data using the coding $y = \dfrac{x-3}{7}$.

b Calculate the mean of the coded data values.

c Use your answer to part **b** to calculate the mean of the original data.

(E) 3 The coded mean price of televisions in a shop was worked out. Using the coding $y = \dfrac{x-65}{200}$ the mean price was 1.5. Find the true mean price of the televisions. **(2 marks)**

4 The coding $y = x - 40$ gives a standard deviation for y of 2.34.
Write down the standard deviation of x.

Watch out Adding or subtracting constants does not affect how spread out the data is, so you can ignore the '−40' when finding the standard deviation for x.

(P) 5 The lifetime, x, in hours, of 70 light bulbs is shown in the table.

Lifetime, x (hours)	$20 < x \leqslant 22$	$22 < x \leqslant 24$	$24 < x \leqslant 26$	$26 < x \leqslant 28$	$28 < x \leqslant 30$
Frequency	3	12	40	10	5

The data is coded using $y = \dfrac{x-1}{20}$.

a Estimate the mean of the coded values $\bar{y}$.

b Hence find an estimate for the mean lifetime of the light bulbs, $\bar{x}$.

c Estimate the standard deviation of the lifetimes of the bulbs.

Problem-solving

Code the midpoints of each class interval. The midpoint of the $22 < x \leqslant 24$ class interval is 23, so the coded midpoint will be $\dfrac{23-1}{20} = 1.1$.

(E) 6 The weekly income, i, of 100 women workers was recorded.

The data was coded using $y = \dfrac{i-90}{100}$ and the following summations were obtained:

    $\Sigma y = 131$, $\Sigma y^2 = 176.84$

Estimate the standard deviation of the actual women workers' weekly income. **(2 marks)**

(E) 7 A meteorologist collected data on the annual rainfall, x mm, at six randomly selected places.

The data was coded using $s = 0.01x - 10$ and the following summations were obtained:

    $\Sigma s = 16.1$, $\Sigma s^2 = 147.03$

Work out an estimate for the standard deviation of the actual annual rainfall. **(2 marks)**

(E/P) 8 A teacher standardises the test marks of his class by adding 12 to each one and then reducing the mark by 20%.

If the standardised marks are represented by t and the original marks by m:

a write down a formula for the coding the teacher has used. **(1 mark)**

The following summary statistics are calculated for the standardised marks:

    $n = 28$ $\bar{t} = 52.8$ $S_{tt} = 7.3$

b Calculate the mean and standard deviation of the original marks gained. **(3 marks)**

9 From the large data set, the daily mean pressure, p hPa, in Hurn during June 2015 is recorded.
The data is coded using $c = \dfrac{p}{2} - 500$ and the following summary statistics are obtained:

$$n = 30 \qquad \bar{c} = 10.15 \qquad S_{cc} = 296.4$$

Find the mean and standard deviation of the daily mean pressure. **(4 marks)**

Mixed exercise 2

1 The mean science mark for one group of eight students is 65. The mean mark for a second group of 12 students is 72. Calculate the mean mark for the combined group of 20 students.

2 The data shows the prices (x) of six shares on a particular day in the year 2007:

$$807 \qquad 967 \qquad 727 \qquad 167 \qquad 207 \qquad 767$$

    a Code the data using the coding $y = \dfrac{x - 7}{80}$.

    b Calculate the mean of the coded data values.

    c Use your answer to part **b** to calculate the mean of the original data.

3 The coded mean of employees' annual earnings (£ x) for a store is 18. The coding used was $y = \dfrac{x - 720}{1000}$. Work out the uncoded mean earnings.

4 Different teachers using different methods taught two groups of students. Both groups of students sat the same examination at the end of the course. The students' marks are shown in the grouped frequency table.

Exam mark	20–29	30–39	40–49	50–59	60–69	70–79	80–89
Frequency group A	1	3	6	6	11	10	8
Frequency group B	1	2	4	13	15	6	3

    a Work out an estimate of the mean mark for group A and an estimate of the mean mark for group B.

    b Write down whether or not the answer to **a** suggests that one method of teaching is better than the other. Give a reason for your answer.

5 The lifetimes of 80 batteries, to the nearest hour, are shown in the table below.

Lifetime (hours)	6–10	11–15	16–20	21–25	26–30
Frequency	2	10	18	45	5

    a Write down the modal class for the lifetime of the batteries.

    b Use interpolation to find the median lifetime of the batteries.

    The midpoint of each class is represented by x and its corresponding frequency by f, giving $\Sigma fx = 1645$.

    c Calculate an estimate of the mean lifetime of the batteries.

    Another batch of 12 batteries is found to have an estimated mean lifetime of 22.3 hours.

    d Estimate the mean lifetime for all 92 batteries.

6 A frequency distribution is shown below.

Class interval	1–20	21–40	41–60	61–80	81–100
Frequency	5	10	15	12	8

Use interpolation to find an estimate for the interquartile range.

7 A frequency distribution is shown below.

Class interval	1–10	11–20	21–30	31–40	41–50
Frequency	10	20	30	24	16

  a Use interpolation to estimate the value of the 30th percentile.

  b Use interpolation to estimate the value of the 70th percentile.

  c Hence estimate the 30% to 70% interpercentile range.

(E) **8** The times it took a random sample of runners to complete a race are summarised in the table.

Time taken (t minutes)	20–29	30–39	40–49	50–59	60–69
Frequency	5	10	36	20	9

  a Use interpolation to estimate the interquartile range. **(3 marks)**

The midpoint of each class was represented by x and its corresponding frequency by f giving:

$$\Sigma fx = 3740 \qquad \Sigma fx^2 = 183\,040$$

  b Estimate the variance and standard deviation for this data. **(3 marks)**

9 The heights of 50 clover flowers are summarised in the table.

Heights in mm (x)	$90 \leqslant x < 95$	$95 \leqslant x < 100$	$100 \leqslant x < 105$	$105 \leqslant x < 110$	$110 \leqslant x < 115$
Frequency	5	10	26	8	1

  a Find Q_1.   **b** Find Q_2.   **c** Find the interquartile range.

  d Use $\Sigma fx = 5075$ and $\Sigma fx^2 = 516\,112.5$ to find the standard deviation.

(E/P) **10** The daily mean temperature is recorded in Camborne during September 2015.

Temperature, t (°C)	$11 < t \leqslant 13$	$13 < t \leqslant 15$	$15 < t \leqslant 17$
Frequency	12	14	4

© Crown Copyright Met Office

  a Use your calculator to find estimates for the mean and standard deviation of the temperatures. **(3 marks)**

  b Use linear interpolation to find an estimate for the 10% to 90% interpercentile range. **(3 marks)**

  c Estimate the number of days in September 2015 where the daily mean temperature in Camborne is more than one standard deviation greater than the mean. **(2 marks)**

(E) **11** The daily mean windspeed, w knots was recorded at Heathrow during May 2015. The data were coded using $z = \dfrac{w - 3}{2}$.

Summary statistics were calculated for the coded data:

$$n = 31 \qquad \Sigma z = 106 \qquad S_{zz} = 80.55$$

    a Find the mean and standard deviation of the coded data. **(2 marks)**

    b Work out the mean and standard deviation of the daily mean windspeed at Heathrow during May 2015. **(2 marks)**

(E) **12** 20 endangered forest owlets were caught for ringing. Their wingspans (x cm) were measured to the nearest centimetre.

The following summary statistics were worked out:

$$\Sigma x = 316 \qquad \Sigma x^2 = 5078$$

    a Work out the mean and the standard deviation of the wingspans of the 20 birds. **(3 marks)**

One more bird was caught. It had a wingspan of 13 centimetres.

    b Without doing any further calculation, say how you think this extra wingspan will affect the mean wingspan. **(1 mark)**

20 giant ibises were also caught for ringing. Their wingspans (y cm) were also measured to the nearest centimetre and the data coded using $z = \dfrac{y - 5}{10}$.

The following summary statistics were obtained from the coded data:

$$\Sigma z = 104 \qquad S_{zz} = 1.8$$

    c Work out the mean and standard deviation of the wingspans of the giant ibis. **(5 marks)**

Challenge

A biologist recorded the heights, x cm, of 20 plant seedlings. She calculated the mean and standard deviation of her results:

$$\bar{x} = 3.1 \text{ cm} \qquad \sigma = 1.4 \text{ cm}$$

The biologist subsequently discovered she had written down one value incorrectly. She replaced a value of 2.3 cm with a value of 3.2 cm.

Calculate the new mean and standard deviation of her data.

Summary of key points

1 The **mode** or **modal class** is the value or class that occurs most often.

2 The **median** is the middle value when the data values are put in order.

3 The **mean** can be calculated using the formula $\bar{x} = \dfrac{\Sigma x}{n}$.

4 For data given in a frequency table, the mean can be calculated using the formula $\bar{x} = \dfrac{\Sigma xf}{\Sigma f}$.

5 To find the **lower quartile** for discrete data, divide n by 4. If this is a whole number, the lower quartile is halfway between this data point and the one above. If it is not a whole number, round *up* and pick this data point.

6 To find the **upper quartile** for discrete data, find $\frac{3}{4}$ of n. If this is a whole number, the upper quartile is halfway between this data point and the one above. If it is not a whole number, round *up* and pick this data point.

7 The **range** is the difference between the largest and smallest values in the data set.

8 The **interquartile range** (IQR) is the difference between the upper quartile and the lower quartile, $Q_3 - Q_1$.

9 The **interpercentile range** is the difference between the values for two given percentiles.

10 **Variance** $= \dfrac{\Sigma(x - \bar{x})^2}{n} = \dfrac{\Sigma x^2}{n} - \left(\dfrac{\Sigma x}{n}\right)^2 = \dfrac{S_{xx}}{n}$ where $S_{xx} = \Sigma(x - \bar{x})^2 = \Sigma x^2 - \dfrac{(\Sigma x)^2}{n}$

11 The **standard deviation** is the square root of the variance:

$$\sigma = \sqrt{\dfrac{\Sigma(x - \bar{x})^2}{n}} = \sqrt{\dfrac{\Sigma x^2}{n} - \left(\dfrac{\Sigma x}{n}\right)^2} = \sqrt{\dfrac{S_{xx}}{n}}$$

12 You can use these versions of the formulae for variance and standard deviation for grouped data that is presented in a frequency table:

$$\sigma^2 = \dfrac{\Sigma f(x - \bar{x})^2}{\Sigma f} = \dfrac{\Sigma fx^2}{\Sigma f} - \left(\dfrac{\Sigma fx}{\Sigma f}\right)^2 \qquad \sigma = \sqrt{\dfrac{\Sigma f(x - \bar{x})^2}{\Sigma f}} = \sqrt{\dfrac{\Sigma fx^2}{\Sigma f} - \left(\dfrac{\Sigma fx}{\Sigma f}\right)^2}$$

where f is the frequency for each group and Σf is the total frequency.

13 If data is coded using the formula $y = \dfrac{x - a}{b}$

- the mean of the coded data is given by $\bar{y} = \dfrac{\bar{x} - a}{b}$

- the standard deviation of the coded data is given by $\sigma_y = \dfrac{\sigma_x}{b}$ where σ_x is the standard deviation of the original data.

3

Representations of data

Prior knowledge check

1 The table shows the number of siblings for 50 year 12 students:

Number of siblings	Frequency
0	5
1	8
2	24
3	10
4	3

a Draw a bar chart to show the data.

b Draw a pie chart to show the data.

← GCSE Mathematics

2 Work out the interquartile range for this set of data:

3, 5, 8, 8, 9, 11, 14, 15, 18, 20, 21, 24

← Section 2.3

3 Work out the mean and standard deviation for this set of data:

17, 19, 20, 25, 28, 31, 32, 32, 35, 37, 38

← Sections 2.1, 2.4

Visual representations can help to illustrate the key features of a data set without the need for complicated calculations. Graphs and charts appear all the time in newspapers and magazines, often stylised to suit the nature of the article.

3.1 Outliers

An outlier is an extreme value that lies outside the overall pattern of the data.

There are a number of different ways of calculating outliers, depending on the nature of the data and the calculations that you are asked to carry out.

- **A common definition of an outlier is any value that is:**
  - **either greater than $Q_3 + k(Q_3 - Q_1)$**
  - **or less than $Q_1 - k(Q_3 - Q_1)$**

Notation Q_1 and Q_3 are the first and third quartiles.

In the exam, you will be told which method to use to identify outliers in data sets, including the value of k.

Example 1

The blood glucose of 30 females is recorded. The results, in mmol/litre, are shown below:

1.7, 2.2, 2.3, 2.3, 2.5, 2.7, 3.1, 3.2, 3.6, 3.7, 3.7, 3.7, 3.8, 3.8, 3.8,

3.8, 3.9, 3.9, 3.9, 4.0, 4.0, 4.0, 4.0, 4.4, 4.5, 4.6, 4.7, 4.8, 5.0, 5.1

An outlier is an observation that falls either 1.5 × interquartile range above the upper quartile or 1.5 × interquartile range below the lower quartile.

a Find the quartiles.   **b** Find any outliers.

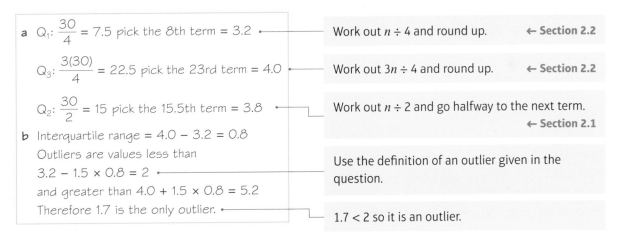

a Q_1: $\dfrac{30}{4} = 7.5$ pick the 8th term = 3.2 — Work out $n \div 4$ and round up. ← **Section 2.2**

Q_3: $\dfrac{3(30)}{4} = 22.5$ pick the 23rd term = 4.0 — Work out $3n \div 4$ and round up. ← **Section 2.2**

Q_2: $\dfrac{30}{2} = 15$ pick the 15.5th term = 3.8 — Work out $n \div 2$ and go halfway to the next term. ← **Section 2.1**

b Interquartile range = 4.0 − 3.2 = 0.8

Outliers are values less than

3.2 − 1.5 × 0.8 = 2 — Use the definition of an outlier given in the question.

and greater than 4.0 + 1.5 × 0.8 = 5.2

Therefore 1.7 is the only outlier. — 1.7 < 2 so it is an outlier.

Example 2

The lengths, in cm, of 12 giant African land snails are given below:

17, 18, 18, 19, 20, 20, 20, 20, 21, 23, 24, 32

a Calculate the mean and standard deviation, given that $\Sigma x = 252$ and $\Sigma x^2 = 5468$.

b An outlier is an observation which lies ±2 standard deviations from the mean. Identify any outliers for this data.

Notation Σx is the sum of the data and Σx^2 is the sum of the square of each value.

a Mean $= \dfrac{\Sigma x}{n} = \dfrac{252}{12} = 21\,\text{cm}$

> Use the summary statistics given to work out the mean and standard deviation quickly.

Variance $= \dfrac{\Sigma x^2}{n} - \bar{x}^2 = \dfrac{5468}{12} - 21^2$

$= 14.666\ldots$

Standard deviation $= \sqrt{14.666\ldots}$

$= 3.83\ (3\ \text{s.f.})$

b Mean $- 2 \times$ standard deviation

$= 21 - 2 \times 3.83 = 13.34$

> Use the definition of an outlier given in the question.

Mean $+ 2 \times$ standard deviation

$= 21 + 2 \times 3.83 = 28.66$

$32\,\text{cm}$ is an outlier.

> **Watch out** Different questions might use different definitions of outliers. Read the question carefully before finding any outliers.

Sometimes outliers are legitimate values which could still be correct. For example, there really could be a giant African land snail 32 cm long.

However, there are occasions when an outlier should be removed from the data since it is clearly an error and it would be misleading to keep it in. These data values are known as **anomalies**.

- **The process of removing anomalies from a data set is known as cleaning the data.**

Anomalies can be the result of experimental or recording error, or could be data values which are not relevant to the investigation.

> **Watch out** Be careful not to remove data values just because they do not fit the pattern of the data. You must justify why a value is being removed.

Here is an example where there is a clear anomaly:

Ages of people at a birthday party: 12, 17, 21, 33, 34, 37, 42, 62, 165

$\bar{x} = 47$ $\qquad \sigma = 44.02 \qquad \bar{x} + 2\sigma = 135.04$

The data value recorded as 165 is significantly higher than $\bar{x} + 2\sigma$, so it can be considered an outlier. An age of 165 is impossible, so this value must be an error. You can clean the data by removing this value before carrying out any analysis.

> **Notation** You can write $165 \gg 135.04$ where $\gg$ is used to denote 'much greater than'. Similarly you can use $\ll$ to denote 'much less than'.

Exercise 3A

1 Some data is collected. $Q_1 = 46$ and $Q_3 = 68$.

A value greater than $Q_3 + 1.5 \times (Q_3 - Q_1)$ or smaller than $Q_1 - 1.5 \times (Q_3 - Q_1)$ is defined as an outlier.

Work out whether the following are outliers using this rule:

a 7 **b** 88 **c** 105

2 The masses of male and female turtles are given in grams. For males, the lower quartile was 400 g and the upper quartile was 580 g. For females, the lower quartile was 260 g and the upper quartile was 340 g.

An outlier is an observation that falls either 1 × (interquartile range) above the upper quartile or 1 × (interquartile range) below the lower quartile.

a Which of these male turtle masses would be outliers?

      400 g   260 g   550 g   640 g

b Which of these female turtle masses would be outliers?

      170 g   300 g   340 g   440 g

c What is the largest mass a male turtle can be without being an outlier?

> **Hint** The definition of an outlier here is different from that in question 1. You will be told which rule to use in the exam.

3 The masses of arctic foxes are found and the mean mass was 6.1 kg. The variance was 4.2.

An outlier is an observation which lies ±2 standard deviations from the mean.

a Which of these arctic fox masses are outliers?

      2.4 kg   10.1 kg   3.7 kg   11.5 kg

b What are the smallest and largest masses that an arctic fox can be without being an outlier?

(E) 4 The ages of nine people at a children's birthday party are recorded. $\Sigma x = 92$ and $\Sigma x^2 = 1428$.

a Calculate the mean and standard deviation of the ages. **(3 marks)**

An outlier is an observation which lies ±2 standard deviations from the mean.

One of the ages is recorded as 30.

b State, with a reason, whether this is an outlier. **(2 marks)**

c Suggest a reason why this age could be a legitimate data value. **(1 mark)**

d Given that all nine people were children, clean the data and recalculate the mean and standard deviation. **(3 marks)**

> **Problem-solving**
> After you clean the data you will need to find the new values for n, Σx and Σx^2.

3.2 Box plots

A box plot can be drawn to represent important features of the data. It shows the quartiles, maximum and minimum values and any outliers.

A box plot looks like this:

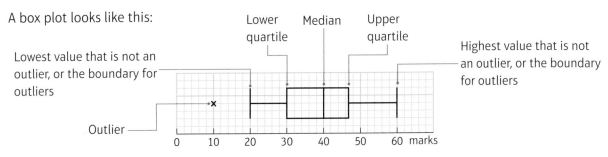

Two sets of data can be compared using box plots.

Example 3

a Draw a box plot for the data on blood glucose levels of females from Example 1.

The blood glucose level of 30 males is recorded. The results, in mmol/litre, are summarised below:

Lower quartile	= 3.6
Upper quartile	= 4.7
Median	= 4.0
Lowest value	= 1.4
Highest value	= 5.2

An outlier is an observation that falls either 1.5 × interquartile range above the upper quartile or 1.5 × interquartile range below the lower quartile.

b Given that there is only one outlier for the males, draw a box plot for this data on the same diagram as the one for females.

c Compare the blood glucose levels for males and females.

a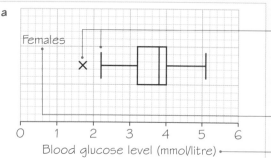

The quartiles and outliers were found in Example 1. ← **page 41**

The outlier is marked with a cross. The lowest value which is not an outlier is 2.2.

Always use a scale and label it. Remember to give your box plot a title.

b Outliers are values less than
$$3.6 - 1.5 \times 1.1 = 1.95$$
and values greater than
$$4.7 + 1.5 \times 1.1 = 6.35$$
There is 1 outlier, which is 1.4.

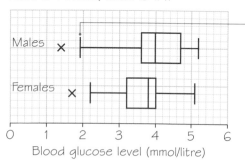

Online Explore box plots and outliers using technology.

The end of the whisker is plotted at the outlier boundary (in this case 1.95) as we do not know the actual figure.

Problem-solving

When drawing two box plots, use the same scale so they can be compared. Remember to give each a title and label the axis.

c The median blood glucose for females is lower than the median blood glucose for males. The interquartile range (the width of the box) and range for blood glucose are smaller for the females.

When comparing data you should compare a measure of location and a measure of spread. You should also write your interpretation in the context of the question.

Exercise 3B

1 A group of students did a test. The summary data is shown in the table.

Lowest value	Lower quartile	Median	Upper quartile	Highest value
5	21	28	36	58

Given that there were no outliers, draw a box plot to illustrate this data.

2 Here is a box plot of marks in an examination.

   a Write down the upper and lower quartiles.

   b Write down the median.

   c Work out the interquartile range.

   d Work out the range.

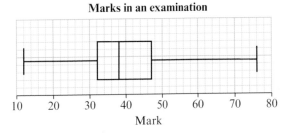

Marks in an examination

Mark

(P) 3 The masses of male and female turtles are given in grams. Their masses are summarised in the box plots.

   a Compare and contrast the masses of the male and female turtles.

   b A turtle was found to have a mass of 330 grams. State whether it is likely to be a male or a female. Give a reason for your answer.

   c Write down the size of the largest female turtle.

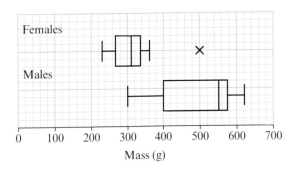

Mass (g)

(E) 4 Data for the maximum daily gust (in knots) in Camborne in September 1987 is taken from the large data set:

13	17	19	20	21
21	22	23	24	25
25	25	26	26	26
27	29	30	30	30
33	35	38	46	78

© Crown Copyright Met Office

   a Calculate Q_1, Q_2 and Q_3. **(3 marks)**

An outlier is defined as a value which lies either 1.5 × the interquartile range above the upper quartile or 1.5 × the interquartile range below the lower quartile.

   b Show that 46 and 78 are outliers. **(1 mark)**

   c Draw a box plot for this data. **(3 marks)**

3.3 Cumulative frequency

If you are given data in a grouped frequency table, you are not able to find the exact values of the median and quartiles. You can draw a **cumulative frequency diagram** and use it to help find estimates for the median, quartiles and percentiles.

Example 4

The table shows the heights, in metres, of 80 giraffes.

a Draw a cumulative frequency diagram.

b Estimate the median height of the giraffes.

c Estimate the lower quartile and the 90th percentile.

d Draw a box plot to represent this data.

Height, h (m)	Frequency
$4.6 \leqslant h < 4.8$	4
$4.8 \leqslant h < 5.0$	7
$5.0 \leqslant h < 5.2$	15
$5.2 \leqslant h < 5.4$	33
$5.4 \leqslant h < 5.6$	17
$5.6 \leqslant h < 5.8$	4

a Add a column to the table to show the cumulative frequency:

Height, h (m)	Frequency	Cumulative frequency
$4.6 \leqslant h < 4.8$	4	4
$4.8 \leqslant h < 5.0$	7	11
$5.0 \leqslant h < 5.2$	15	26
$5.2 \leqslant h < 5.4$	33	59
$5.4 \leqslant h < 5.6$	17	76
$5.6 \leqslant h < 5.8$	4	80

$4 + 7 = 11$

$11 + 15 = 26$

This represents the number of data values that are in the range $4.6 \leqslant h < 5.4$.

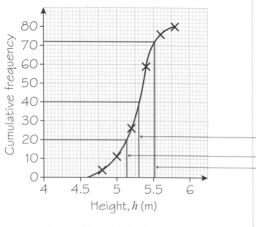

The lowest possible value for the height is 4.6 m so plot (4.6, 0).

Plot each point using the upper class boundary for x and the cumulative frequency for y: coordinates (4.8, 4), (5.0, 11), (5.2, 26), (5.4, 59), (5.6, 76) and (5.8, 80). Join the points with a smooth curve.

For part **b**, draw a line across from 40 on the cumulative frequency axis and then down to the height axis.

b The median is the 40th data point.
An estimate for the median is 5.3 m.

c The lower quartile is the 20th data point.
The 90th percentile is the 72nd data point.
An estimate for the lower quartile is 5.15 m.
An estimate for the 90th percentile is 5.52 m.

For part **c**, draw lines to estimate the lower quartile and the 90th percentile.

The data is continuous so you do not add 1.

← Section 2.2

d   **Height of giraffes**

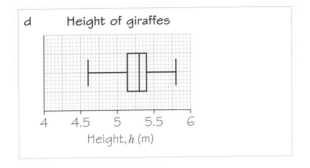

Height, h (m)

The minimum possible data value is 4.6 and the maximum possible data value is 5.8. The median and lower quartile are taken from parts **b** and **c**. The upper quartile is at 5.4.

Exercise 3C

1 The table shows the masses, in kilograms, of 120 Coulter pine cones.

a Draw a cumulative frequency diagram for this data.

b Estimate the median mass.

c Find the interquartile range and the 10th to 90th interpercentile range.

d Draw a box plot to show this data.

Mass, m (kg)	Frequency
$1.0 \leqslant m < 1.2$	7
$1.2 \leqslant m < 1.4$	18
$1.4 \leqslant m < 1.6$	34
$1.6 \leqslant m < 1.8$	41
$1.8 \leqslant m < 2.0$	15
$2.0 \leqslant m < 2.2$	5

2 The table shows the lengths, in cm, of 70 earthworms.

a Draw a cumulative frequency diagram for this data.

b Estimate the median and quartiles.

c Estimate how many earthworms are
  i longer than 8.2 cm   **ii** shorter than 7.3 cm.

d Draw a box plot to show this data.

Length, l (cm)	Frequency
$6.0 \leqslant l < 6.5$	3
$6.5 \leqslant l < 7.0$	13
$7.0 \leqslant l < 7.5$	14
$7.5 \leqslant l < 8.0$	26
$8.0 \leqslant l < 8.5$	10
$8.5 \leqslant l < 9.0$	4

(P) 3 The table shows the times taken by 80 men and 80 women to complete a crossword puzzle.

a Draw cumulative frequency diagrams for both sets of data on the same axes.

b Which gender had the lower median time?

c Which gender had the bigger spread of times?

d The qualifying time for the next round of a national competition is 7.5 minutes. Estimate the numbers of men and women who qualified for the competition.

Time, t (min)	Frequency (men)	Frequency (women)
$5 \leqslant t < 6$	2	3
$6 \leqslant t < 7$	14	15
$7 \leqslant t < 8$	17	21
$8 \leqslant t < 9$	40	35
$9 \leqslant t < 10$	7	6

Problem-solving

To compare spread you could use the interquartile range or the 10th to 90th percentile range.

4 A vet measures the masses, in kg, of male and female domestic shorthair cats. Her results are given in the table.

a Draw cumulative frequency diagrams for both sets of data on the same axes.

b Which gender has the greater spread of masses?

A female domestic shorthair cat is considered underweight if its mass is below 3.2 kg.

A male domestic shorthair cat is considered underweight if its mass is below 3.8 kg.

c Which gender has fewer underweight cats?

Mass, w (kg)	Frequency (male)	Frequency (female)
$2.5 \leqslant w < 3.0$	1	5
$3.0 \leqslant w < 3.5$	12	17
$3.5 \leqslant w < 4.0$	20	32
$4.0 \leqslant w < 4.5$	27	12
$4.5 \leqslant w < 5.0$	7	4
$5.0 \leqslant w < 5.5$	3	0

3.4 Histograms

Grouped continuous data can be represented in a **histogram**.

Generally, a histogram gives a good picture of how the data is distributed. It enables you to see a rough location, the general shape and how spread out the data is.

In a histogram, the **area** of the bar is proportional to the frequency in each class. This allows you to use a histogram to represent grouped data with unequal class intervals.

- **On a histogram, to calculate the height of each bar (the frequency density) use the formula**

  area of bar $= k \times$ frequency.

  $k = 1$ is the easiest value to use when drawing a histogram.

  If $k = 1$, then

  $$\text{frequency density} = \frac{\text{frequency}}{\text{class width}}$$

- **Joining the middle of the top of each bar in a histogram forms a frequency polygon.**

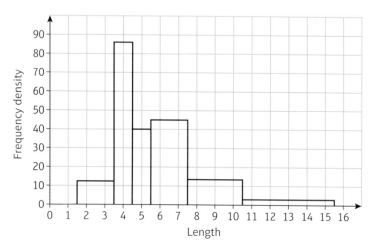

Example 5

A random sample of 200 students was asked how long it took them to complete their homework the previous night. The time was recorded and summarised in the table below.

Time, t (minutes)	$25 \leqslant t < 30$	$30 \leqslant t < 35$	$35 \leqslant t < 40$	$40 \leqslant t < 50$	$50 \leqslant t < 80$
Frequency	55	39	68	32	6

a Draw a histogram and a frequency polygon to represent the data.

b Estimate how many students took between 36 and 45 minutes to complete their homework.

	Time, t (minutes)	Frequency	Class width	Frequency density
a	$25 \leqslant t < 30$	55	5	11
	$30 \leqslant t < 35$	39	5	7.8
	$35 \leqslant t < 40$	68	5	13.6
	$40 \leqslant t < 50$	32	10	3.2
	$50 \leqslant t < 80$	6	30	0.2

Frequency density $= \frac{55}{5} = 11$

Class width $= 30 - 25 = 5$

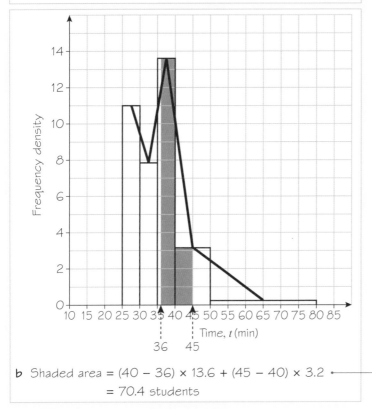

To draw the frequency polygon, join the middle of the top of each bar of the histogram.

b Shaded area $= (40 - 36) \times 13.6 + (45 - 40) \times 3.2$
$= 70.4$ students

To estimate the number of students who spent between 36 and 45 minutes, you need to find the area between 36 and 45.

Example 6

A random sample of daily mean temperatures (T, °C) was taken from the large data set for Hurn in 2015. The temperatures were summarised in a grouped frequency table and represented by a histogram.

a Give a reason to support the use of a histogram to represent this data.

b Write down the underlying feature associated with each of the bars in a histogram.

On the histogram the rectangle representing the $16 \leqslant T < 18$ class was 3.2 cm high and 2 cm wide. The frequency for this class was 8.

c Show that each day is represented by an area of 0.8 cm².

d Given that the total area under the histogram was 48 cm², find the total number of days in the sample.

a Temperature is continuous and the data were given in a grouped frequency table.

b The area of the bar is proportional to the frequency.

c Area of bar = 3.2 × 2 = 6.4 ────── An area of 6.4 cm² represents a frequency of 8.
6.4 ÷ 8 = 0.8 cm²

d There were 60 days in the sample. ────── Total number of days × area for one day = total area under histogram

Exercise 3D

1 The data shows the mass, in pounds, of 50 adult puffer fish.

  a Draw a histogram for this data.

  b On the same set of axes, draw a frequency polygon.

Mass, m (pounds)	Frequency
$10 \leqslant m < 15$	4
$15 \leqslant m < 20$	12
$20 \leqslant m < 25$	23
$25 \leqslant m < 30$	8
$30 \leqslant m < 35$	3

(P) 2 Some students take part in an obstacle race. The time it took each student to complete the race was noted. The results are shown in the histogram.

  a Give a reason to justify the use of a histogram to represent this data.

The number of students who took between 60 and 70 seconds is 90.

  b Find the number of students who took between 40 and 60 seconds.

  c Find the number of students who took 80 seconds or less.

  d Calculate the total number of students who took part in the race.

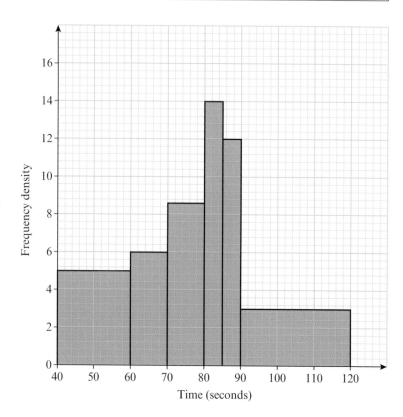

Watch out Frequency density × class width is always **proportional** to frequency in a histogram, but not necessarily **equal** to frequency.

(P) **3** A Fun Day committee at a local sports centre organised a throwing the cricket ball competition. The distance thrown by every competitor was recorded. The histogram shows the data. The number of competitors who threw less than 20 m was 40.

a Why is a histogram a suitable diagram to represent this data?

b How many people entered the competition?

c Estimate how many people threw between 30 and 40 metres.

d How many people threw between 45 and 65 metres?

e Estimate how many people threw less than 25 metres.

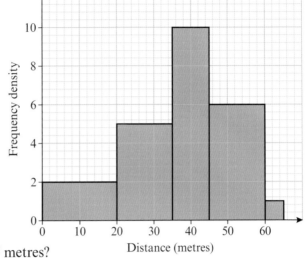

(P) **4** A farmer found the masses of a random sample of lambs. The masses were summarised in a grouped frequency table and represented by a histogram. The frequency for the class $28 \leqslant m < 32$ was 32.

a Show that 25 small squares on the histogram represents 8 lambs.

b Find the frequency of the $24 \leqslant m < 26$ class.

c How many lambs did the farmer weigh in total?

d Estimate the number of lambs that had masses between 25 and 29 kg.

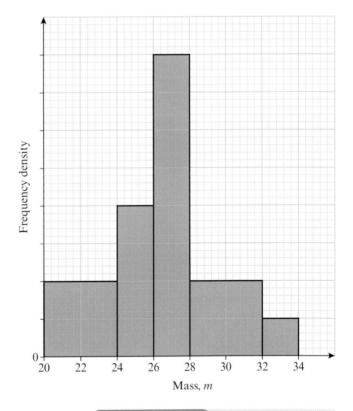

Problem-solving

You can use area to solve histogram problems where no vertical scale is given. You could also use the information given in the question to work out a suitable scale for the vertical axis.

E/P **5** The partially completed histogram
shows the time, in minutes, that
passengers were delayed at an airport.

a i Copy and complete the table.

Time, t (min)	Frequency
$0 \leqslant t < 20$	4
$20 \leqslant t < 30$	
$30 \leqslant t < 35$	15
$35 \leqslant t < 40$	25
$40 \leqslant t < 50$	
$50 \leqslant t < 70$	

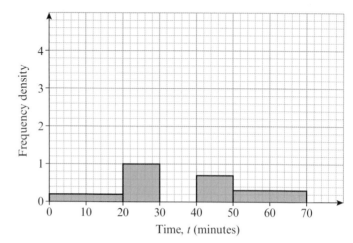

ii Copy and complete the histogram. **(4 marks)**

b Estimate the number of passengers that were delayed for between 25 and 38 minutes. **(2 marks)**

E/P **6** The variable y was measured to the nearest whole number. 60 observations were taken and are
recorded in the table below.

y	10–12	13–14	15–17	18–25
Frequency	6	24	18	12

a Write down the class boundaries for the 13–14 class. **(1 mark)**

A histogram was drawn and the bar representing the 13–14 class had a width of 4 cm and a
height of 6 cm.

For the bar representing the 15–17 class, find:

Problem-solving

Remember that area is proportional to frequency.

b i the width **(1 mark)**

  ii the height. **(2 marks)**

E/P **7** From the large data set, the daily mean temperature
for Leeming during May 2015 is summarised in the table.

A histogram was drawn. The $8 \leqslant t < 10$ group was
represented by a bar of width 1 cm and a height of 8 cm.

a Find the width and height of the bar representing the
$11 \leqslant t < 12$ group. **(2 marks)**

b Use your calculator to estimate the mean and
standard deviation of temperatures in Leeming
in May 2015. **(3 marks)**

Daily mean temperature, t (°C)	Frequency
$4 \leqslant t < 8$	4
$8 \leqslant t < 10$	8
$10 \leqslant t < 11$	6
$11 \leqslant t < 12$	7
$12 \leqslant t < 15$	5
$15 \leqslant t < 16$	1

© Crown Copyright Met Office

c Use linear interpolation to find an estimate for the lower quartile of temperatures. **(2 marks)**

d Estimate the number of days in May 2015 on which the temperature was higher than
the mean plus one standard deviation. **(2 marks)**

3.5 Comparing data

- **When comparing data sets you can comment on:**
  - **a measure of location**
  - **a measure of spread**

You can compare data using the mean and standard deviation or using the median and interquartile range. If the data set contains extreme values, then the median and interquartile range are more appropriate statistics to use.

Watch out Do not use the median with the standard deviation or the mean with the interquartile range.

Example 7

From the large data set, the daily mean temperature during August 2015 is recorded at Heathrow and Leeming.

For Heathrow, $\Sigma x = 562.0$ and $\Sigma x^2 = 10\,301.2$.

a Calculate the mean and standard deviation for Heathrow.

For Leeming, the mean temperature was 15.6 °C with a standard deviation of 2.01 °C.

b Compare the data for the two locations using the information given.

a Mean = 562.0 ÷ 31 = 18.12... = 18.1 °C

Standard deviation

$$= \sqrt{\frac{10\,301.2}{31} - \left(\frac{562.0}{31}\right)^2} = 1.906...$$

$$= 1.91\,°C\ (3\ \text{s.f.})$$

Use $\bar{x} = \frac{\Sigma x}{n}$. There are 31 days in August so $n = 31$.

Use your calculator to do this calculation in one step. Round your final answer to 3 significant figures.

b The mean daily temperature in Leeming is lower than in Heathrow and the spread of temperatures is greater than in Heathrow.

Compare the mean and standard deviation as a measure of location and a measure of spread.

Exercise 3E

Speeds of cars on motorways

(P) 1 The box plots show the distribution of speeds of cars on two motorways.

Compare the distributions of the speeds on the two motorways.

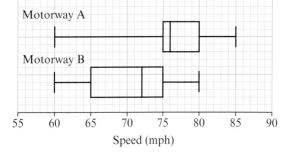

(P) 2 Two classes of primary school children complete a puzzle. Summary statistics for the times, in minutes, the children took are shown in the table.

Calculate the mean and standard deviation of the times and compare the distributions.

	n	Σx	Σx^2
Class 2B	20	650	22 000
Class 2F	22	598	19 100

P **3** The cumulative frequency diagram
shows the distribution of heights of
80 boys and 80 girls in a basketball club.

Compare the heights of boys and girls
in the club.

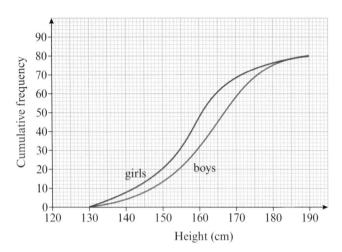

E **4** A sample of the daily maximum relative humidity is taken from the large data set for Leuchars
and for Camborne during 2015. The data is given in the table.

Leuchars	100	98	100	100	100	100	100	100	94	100	91	100	100	89	100
Camborne	92	95	99	96	100	100	90	98	81	99	100	99	91	98	100

© Crown Copyright Met Office

a Find the median and quartiles for both samples. **(4 marks)**

b Compare the two samples. **(2 marks)**

Large data set

You will need access to the large data set and spreadsheet software to
answer this question. Look at the data on daily mean windspeed in Hurn
in 1987 and in 2015.

1 For each year, calculate:

   a the mean   **b** the mode   **c** the standard deviation.

2 Compare the daily mean windspeeds in Hurn in 1987 and 2015.

Hint You can use the MODE
function in a spreadsheet to
find the modal value from a
range of cells.

Mixed exercise **3**

1 Jason and Perdita decided to go on a touring holiday on the continent for the whole of July.
They recorded the distance they travelled, in kilometres, each day:

      155, 164, 168, 169, 173, 175, 177, 178, 178, 178, 179, 179, 179, 184, 184, 185,
      185, 188, 192, 193, 194, 195, 195, 196, 204, 207, 208, 209, 211, 212, 226

a Find Q_1, Q_2 and Q_3

Outliers are values that lie outside $Q_1 - 1.5(Q_3 - Q_1)$ and $Q_3 + 1.5(Q_3 - Q_1)$.

b Find any outliers.

c Draw a box plot of this data.

P **2** Fell runners from the Esk Club and the Irt Club were keen to see which club had the faster runners overall. They decided that all the members from both clubs would take part in a fell run. The time each runner took to complete the run was recorded.

The results are summarised in the box plot.

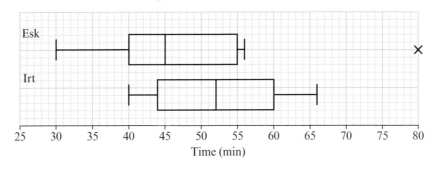

Time (min)

a Write down the time by which 50% of the Esk Club runners had completed the run.

b Write down the time by which 75% of the Irt Club runners had completed the run.

c Explain what is meant by the cross (×) on the Esk Club box plot.

d Compare and contrast these two box plots.

e What conclusions can you draw from this information about which club has the faster runners?

f Give one advantage and one disadvantage of comparing distributions using box plots.

P **3** The table shows the lengths, in cm, of 60 honey badgers.

a Draw a cumulative frequency diagram for this data.

b Find the median length of a honey badger.

c Find the interquartile range.

Length, x (cm)	Frequency
$50 \leqslant x < 55$	2
$55 \leqslant x < 60$	7
$60 \leqslant x < 65$	15
$65 \leqslant x < 70$	31
$70 \leqslant x < 75$	5

This diagram shows the distribution of lengths of European badgers.

d Compare the distributions of lengths of honey badgers and European badgers.

e Comment on the suitability of using cumulative frequency diagrams to compare these distributions.

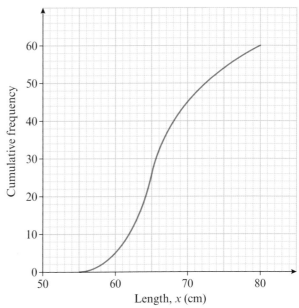

Length, x (cm)

(P) **4** The histogram shows the time taken by a group of 58 girls to run a measured distance.

    a Work out the number of girls who took longer than 56 seconds.

    b Estimate the number of girls who took between 52 and 55 seconds.

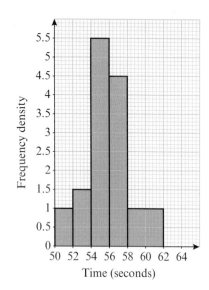

(E/P) **5** The table gives the distances travelled to school, in km, of the population of children in a particular region of the United Kingdom.

Distance, d (km)	$0 \leqslant d < 1$	$1 \leqslant d < 2$	$2 \leqslant d < 3$	$3 \leqslant d < 5$	$5 \leqslant d < 10$	$10 \leqslant d$
Number	2565	1784	1170	756	630	135

A histogram of this data was drawn with distance along the horizontal axis. A bar of horizontal width 1.5 cm and height 5.7 cm represented the 0–1 km group.

Find the widths and heights, in cm, to one decimal place, of the bars representing the following groups:

    a $2 \leqslant d < 3$ **b** $5 \leqslant d < 10$ **(5 marks)**

6 The labelling on bags of garden compost indicates that the bags have a mass of 20 kg. The masses of a random sample of 50 bags are summarised in the table opposite.

Mass, m (kg)	Frequency
$14.6 \leqslant m < 14.8$	1
$14.8 \leqslant m < 18.0$	0
$18.0 \leqslant m < 18.5$	5
$18.5 \leqslant m < 20.0$	6
$20.0 \leqslant m < 20.2$	22
$20.2 \leqslant m < 20.4$	15
$20.4 \leqslant m < 21.0$	1

    a On graph paper, draw a histogram of this data.

    b Estimate the mean and standard deviation of the mass of a bag of compost.

    [You may use $\Sigma fy = 988.85$, $\Sigma fy^2 = 19\,602.84$]

    c Using linear interpolation, estimate the median.

7 The number of bags of potato crisps sold per day in a bar was recorded over a two-week period. The results are shown below.

       20   15   10   30   33   40   5   11   13   20   25   42   31   17

    a Calculate the mean of this data.

    b Find the median and the quartiles of this data.

An outlier is an observation that falls either 1.5 × (interquartile range) above the upper quartile or 1.5 × (interquartile range) below the lower quartile.

    c Determine whether or not any items of data are outliers.

    d On graph paper draw a box plot to represent this data. Show your scale clearly.

E **8** From the large data set, the daily maximum gust (knots) is measured at Hurn throughout May and June 2015. The data is summarised in the table.

A histogram is drawn to represent this data. The bar representing the $10 \leqslant g < 15$ class is 2.5 cm wide and 1.8 cm high.

Daily maximum gust, g (knots)	Frequency
$10 \leqslant g < 15$	3
$15 \leqslant g < 18$	9
$18 \leqslant g < 20$	9
$20 \leqslant g < 25$	20
$25 \leqslant g < 30$	9
$30 \leqslant g < 50$	7

© Crown Copyright Met Office

a Give a reason to support the use of a histogram to represent this data. **(1 mark)**

b Calculate the width and height of the bar representing the $18 \leqslant g < 20$ class. **(3 marks)**

c Use your calculator to estimate the mean and standard deviation of the maximum gusts. **(3 marks)**

d Use linear interpolation to find an estimate for the number of days the maximum gust was within one standard deviation of the mean. **(4 marks)**

E **9** From the large data set, data was gathered in September 1987 and in September 2015 for the mean daily temperature in Leuchars. Summary statistics are given in the table.

	Min	Max	Median	Σx	Σx^2
1987	7.0	17.0	11.85	356.1	4408.9
2015	10.1	14.1	12.0	364.1	4450.2

a Calculate the mean of the mean daily temperatures in each of the two years. **(2 marks)**

b In 2015, the standard deviation was 1.02. Compare the mean daily temperatures in the two years. **(2 marks)**

c A recorded temperature is considered 'normal' for the time of year if it is within one standard deviation of the mean. Estimate for how many days in September 2015 a 'normal' mean daily temperature was recorded. State one assumption you have made in making the estimate. **(3 marks)**

Challenge

The table shows the lengths of the films in a film festival, to the nearest minute.

Length (min)	Frequency
70–89	4
90–99	17
100–109	20
110–139	9
140–179	2

A histogram is drawn to represent the data, and the bar representing the 90–99 class is 3 cm higher than the bar representing the 70–89 class.

Find the height of the bar representing the 110–139 class.

Summary of key points

1 A common definition of an outlier is any value that is:
- either greater than $Q_3 + k(Q_3 - Q_1)$
- or less than $Q_1 - k(Q_3 - Q_1)$

2 The process of removing anomalies from a data set is known as cleaning the data.

3 On a histogram, to calculate the height of each bar (the **frequency density**) use the formula area of bar = $k \times$ frequency.

4 Joining the middle of the top of each bar in a histogram forms a frequency polygon.

5 When comparing data sets you can comment on:
- a measure of location
- a measure of spread

Correlation

4

Objectives

After completing this chapter you should be able to:

* Draw and interpret scatter diagrams for bivariate data → pages 60–61

* Interpret correlation and understand that it does not imply causation → pages 61–62

* Interpret the coefficients of a regression line equation for bivariate data → pages 63–64

* Understand when you can use a regression line to make predictions → pages 64–66

Prior knowledge check

1. The table shows the scores out of 10 on a maths test and on a physics test for 7 students.

Maths	6	7	7	8	9	9	10
Physics	9	7	6	7	5	4	5

Show this information on a scatter diagram.

← GCSE Mathematics

2. A straight line has equation $y = 0.34 - 3.21x$. Write down

   a the gradient of the line

   b the y-intercept of the line

   ← GCSE Mathematics

Climate scientists have demonstrated a strong correlation between greenhouse gas emissions and rising atmospheric temperatures.

→ Mixed exercise Q2

4.1 Correlation

- **Bivariate data is data which has pairs of values for two variables.**

You can represent bivariate data on a **scatter diagram**. This scatter diagram shows the results from an experiment on how breath rate affects pulse rate:

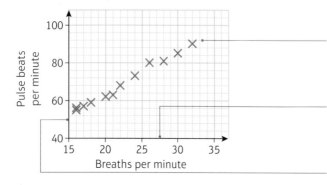

Each cross represents a data point. This subject had a breath rate of 32 breaths per minute and a pulse rate of 89 beats per minute.

The researcher could control this variable. It is called the **independent** or **explanatory variable**. It is usually plotted on the horizontal axis.

The researcher measured this variable. It is called the **dependent** or **response variable**. It is usually plotted on the vertical axis.

The two different variables in a set of bivariate data are often related.

- **Correlation describes the nature of the linear relationship between two variables.**

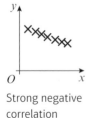

Strong negative correlation

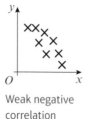

Weak negative correlation

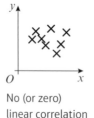

No (or zero) linear correlation

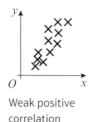

Weak positive correlation

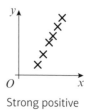

Strong positive correlation

For negatively correlated variables, when one variable increases the other decreases.

For positively correlated variables, when one variable increases, the other also increases.

Watch out You should only use correlation to describe data that shows a linear relationship. Variables with no linear correlation could still show a non-linear relationship.

Example 1

In the study of a city, the population density, in people/hectare, and the distance from the city centre, in km, was investigated by picking a number of sample areas with the following results.

Area	A	B	C	D	E	F	G	H	I	J
Distance (km)	0.6	3.8	2.4	3.0	2.0	1.5	1.8	3.4	4.0	0.9
Population density (people/hectare)	50	22	14	20	33	47	25	8	16	38

a Draw a scatter diagram to represent this data.

b Describe the correlation between distance and population density.

c Interpret your answer to part **b**.

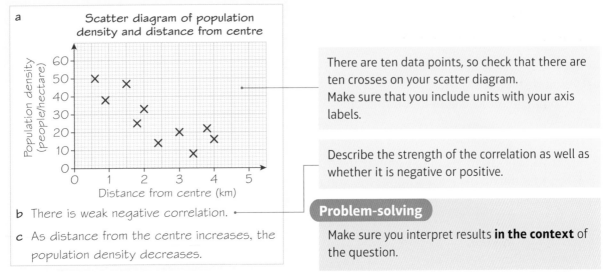

Scatter diagram of population density and distance from centre

a

There are ten data points, so check that there are ten crosses on your scatter diagram.
Make sure that you include units with your axis labels.

Describe the strength of the correlation as well as whether it is negative or positive.

b There is weak negative correlation.

Problem-solving

c As distance from the centre increases, the population density decreases.

Make sure you interpret results **in the context** of the question.

Two variables have a **causal relationship** if a change in one variable causes a change in the other. Just because two variables show correlation it does not necessarily mean that they have a causal relationship.

■ **When two variables are correlated, you need to consider the context of the question and use your common sense to determine whether they have a causal relationship.**

Example 2

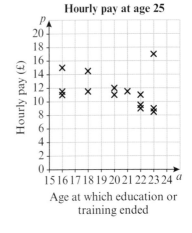

Hideko was interested to see if there was a relationship between what people earn and the age at which they left education or training. She asked 14 friends to fill in an anonymous questionnaire and recorded her results in a scatter diagram.

a Describe the type of correlation shown.

Hideko says that her data supports the conclusion that more education causes people to earn a lower hourly rate of pay.

b Give one reason why Hideko's conclusion might not be valid.

a Weak negative correlation.

b Respondents who left education later would have significantly less work experience than those who left education earlier. This could be the cause of the reduced income shown in her results.

You could also say that Hideko's conclusion is not valid because she used a small, opportunistic sample.   ← **Section 1.1**

Exercise 4A

1 Some research was done into the effectiveness of a weight-reducing drug. Seven people recorded their weight loss and this was compared with the length of time for which they had been treated. A scatter diagram was drawn to represent this data.

a Describe the type of correlation shown by the scatter diagram.

b Interpret the correlation in context.

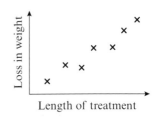

2 The average temperature and rainfall were collected for a
number of cities around the world.
The scatter diagram shows this information.

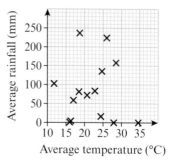

a Describe the correlation between average temperature
and average rainfall.

b Comment on the claim that hotter cities have less rainfall.

3 Eight students were asked to estimate the mass of a bag of sweets in grams. First they were
asked to estimate the mass without touching the bag and then they were told to pick the bag up
and estimate the mass again. The results are shown in the table below.

Student	A	B	C	D	E	F	G	H
Estimate of mass not touching bag (g)	25	18	32	27	21	35	28	30
Estimate of mass holding bag (g)	16	11	20	17	15	26	22	20

a Draw a scatter diagram to represent this data.

b Describe and interpret the correlation between the two variables.

4 Donal was interested to see whether there was a relationship between the value of a house and
the speed of its internet connection, as measured by the time taken to download a 100 megabyte
file. The table shows his results.

Time taken (s)	5.2	5.5	5.8	6.0	6.8	8.3	9.3	13	13.6	16.0
House value (£1000s)	300	310	270	200	230	205	208	235	175	180

a Draw a scatter diagram to represent this data.

b Describe the type of correlation shown.

Donal says that his data shows that a slow internet connection reduces the value of a house.

c Give one reason why Donal's conclusion may not be valid.

(E) 5 The table shows the daily total rainfall, r mm, and daily total hours of sunshine, s, in Leuchars
for a random sample of 11 days in August 1987, from the large data set.

r	0	6.8	0.9	4.8	0	21.7	1.7	4.9	0.1	2.2	0.1
s	8.4	4.9	10.2	4.5	3.3	3.9	5.4	1.8	9.7	1	4.6

© Crown Copyright Met Office

The median and quartiles for the rainfall data are: $Q_1 = 0.1$ $Q_2 = 1.7$ $Q_3 = 4.85$

An outlier is defined as a value which lies either $1.5 \times$ the interquartile range above the upper
quartile or $1.5 \times$ the interquartile range below the lower quartile.

a Show that $r = 21.7$ is an outlier.   **(1 mark)**

b Give a reason why you might:
   i include   ii exclude this day's readings.   **(2 marks)**

c Exclude this day's readings and draw a scatter diagram to represent the data for the
   remaining ten days.   **(3 marks)**

d Describe the correlation between rainfall and hours of sunshine.   **(1 mark)**

e Do you think there is a causal relationship between the amount of rain and the
   hours of sunshine on a particular day? Explain your reasoning.   **(1 mark)**

4.2 Linear regression

When a scatter diagram shows correlation, you can draw a **line of best fit**. This is a linear model that approximates the relationship between the variables. One type of line of best fit that is useful in statistics is a **least squares regression line**. This is the straight line that minimises the sum of the squares of the distances of each data point from the line.

Notation The least squares regression line is usually just called the regression line.

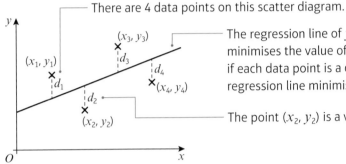

There are 4 data points on this scatter diagram.

The regression line of y on x is the straight line that minimises the value of $d_1^2 + d_2^2 + d_3^2 + d_4^2$. In general, if each data point is a distance d_i from the line, the regression line minimises the value of Σd_i^2.

The point (x_2, y_2) is a vertical distance d_2 from the line.

- **The regression line of y on x is written in the form $y = a + bx$.**

You can use a calculator to find the values of the coefficients a and b for a given set of bivariate data. You will not be required to do this in your exam.

Watch out The order of the variables is important. The regression line of y on x will be different from the regression line of x on y.

- **The coefficient b tells you the change in y for each unit change in x.**
  - **If the data is positively correlated, b will be positive.**
  - **If the data is negatively correlated, b will be negative.**

Example 3

From the large data set, the daily mean windspeed, w knots, and the daily maximum gust, g knots, were recorded for the first 15 days in May in Camborne in 2015.

w	14	13	13	9	18	18	7	15	10	14	11	9	8	10	7
g	33	37	29	23	43	38	17	30	28	29	29	23	21	28	20

© Crown Copyright Met Office

The data was plotted on a scatter diagram:

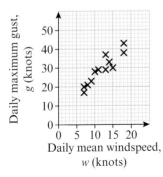

a Describe the correlation between daily mean windspeed and daily maximum gust.

The equation of the regression line of g on w for these 15 days is $g = 7.23 + 1.82w$.

b Give an interpretation of the value of the gradient of this regression line.

c Justify the use of a linear regression line in this instance.

a There is a strong positive correlation between daily mean windspeed and daily maximum gust.

b If the daily mean windspeed increases by 10 knots the daily maximum gust increases by approximately 18 knots.

c The correlation suggests that there is a linear relationship between g and w so a linear regression line is a suitable model.

Online Explore the regression line and analysis using technology.

Make sure your interpretation refers to both the context and your numerical value of the gradient. Try to phrase your answer as a complete, clear sentence.

Problem-solving

A regression line is a valid model when the data shows linear correlation. The stronger the correlation, the more accurately the regression line will model the data.

If you know a value of the **independent variable** from a bivariate data set, you can use the regression line to make a prediction or estimate of the corresponding value of the **dependent variable**.

- **You should only use the regression line to make predictions for values of the dependent variable that are within the range of the given data.**

Notation This is called **interpolation**. Making a prediction based on a value outside the range of the given data is called **extrapolation**, and gives a much less reliable estimate.

Example 4

The head circumference, y cm, and gestation period, x weeks, for a random sample of eight newborn babies at a clinic were recorded.

Gestation period (x weeks)	36	40	33	37	40	39	35	38
Head circumference (y cm)	30.0	35.0	29.8	32.5	33.2	32.1	30.9	33.6

The scatter graph shows the results.

The equation of the regression line of y on x is $y = 8.91 + 0.624x$.

The regression equation is used to estimate the head circumference of a baby born at 39 weeks and a baby born at 30 weeks.

a Comment on the reliability of these estimates.

A nurse wants to estimate the gestation period for a baby born with a head circumference of 31.6 cm.

b Explain why the regression equation given above is not suitable for this estimate.

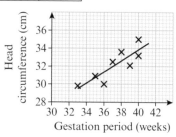

a The prediction for 39 weeks is within the range of the data (interpolation) so is more likely to be accurate.

The prediction for 30 weeks is outside the range of the data (extrapolation) so is less likely to be accurate.

You could also comment on the sample. The sample was randomly chosen which would improve the accuracy of the predictions, but the sample size is small which would reduce the accuracy of the predictions.

b The independent (explanatory) variable in this model is the gestation period, x. You should not use this model to predict a value of x for a given value of y.

Watch out You should only make predictions for the **dependent** variable. If you needed to predict a value of x for a given value of y you would need to use the regression line of x on y.

Exercise 4B

1 An accountant monitors the number of items produced per month by a company together with the total production costs. The table shows this data.

Number of items, n (1000s)	21	39	48	24	72	75	15	35	62	81	12	56
Production costs, p (£1000s)	40	58	67	45	89	96	37	53	83	102	35	75

a Draw a scatter diagram to represent this data.

The equation of the regression line of p on n is $p = 21.0 + 0.98n$.

b Draw the regression line on your scatter diagram.

c Interpret the meaning of the figures 21.0 and 0.98.

The company expects to produce 74 000 items in June, and 95 000 items in July.

d Comment on the suitability of this regression line equation to predict the production costs in each of these months.

2 The relationship between the number of coats of paint applied to a boat and the resulting weather resistance was tested in a laboratory. The data collected is shown in the table.

a Draw a scatter diagram to represent this data.

The equation of the regression line is $y = 2.93 + 1.45x$.

Helen says that a gradient of 1.45 means that if 10 coats of paint are applied the protection will last 14.5 years.

b Comment on Helen's statement.

Coats of paint (x)	Protection (years) (y)
1	4.4
2	5.9
3	7.1
4	8.8
5	10.2

3 The table shows the ages of some chickens and the number of eggs that they laid in a month.

Age of chicken, a (months)	18	32	44	60	71	79	99	109	118	140
Number of eggs laid in a month, n	16	18	13	7	12	7	11	13	6	9

a Draw a scatter diagram to show this information.

Robin calculates the regression line of n on a as $n = 16.1 + 0.063a$.

b Without further calculation, explain why Robin's regression equation is incorrect.

4 Aisha collected data on the numbers of bedrooms, x, and the values, y (£1000s), of the houses in her village. She calculates the regression equation of y on x to be $y = 190 + 50x$.

She states that the value of the constant in her regression equation means that a house with no bedrooms in her village would be worth £190 000. Explain why this is not a reasonable statement.

(E) **5** The table shows the daily maximum relative humidity, h (%), and the daily mean visibility, v decametres (Dm), in Heathrow for the first two weeks in September 2015, from the large data set.

h	94	95	92	80	97	94	93	90	87	95	93	92	91	98
v	2600	2900	3900	4300	2800	2400	2700	3500	3000	2200	2200	3300	2800	2200

© Crown Copyright Met Office

The equation of the regression line of v on h is $v = 12\,700 - 106h$

a Give an interpretation of the value of the gradient of the regression line. **(1 mark)**

b Use your knowledge of the large data set to explain whether there is likely to be a causal relationship between humidity and visibility. **(2 marks)**

c Give reasons why it would not be reliable to use this regression equation to predict:

  i the mean visibility on a day with 100% humidity **(2 marks)**

  ii the humidity on a day with visibility of 3000 dm. **(2 marks)**

d State two ways in which better use could be made of the large data set to produce a model describing the relationship between humidity and visibility. **(2 marks)**

Mixed exercise (4)

1 A survey of British towns recorded the number of serious road accidents in a week (x) in each town, together with the number of fast food restaurants (y). The data showed a strong positive correlation. Katie states that this shows that building more fast food restaurants in her town will cause more serious road accidents. Explain whether the data supports Katie's statement.

2 The following table shows the mean CO_2 concentration in the atmosphere, c (ppm), and the increase in average temperature compared to the 30-year period 1951–1980, t (°C).

Year	2015	2013	2011	2009	2007	2005	2003	2001	1999	1997	1995	1994
c (ppm)	401	397	392	387	384	381	376	371	368	363	361	357
t (°C)	0.86	0.65	0.59	0.64	0.65	0.68	0.61	0.54	0.41	0.47	0.45	0.24

Source: Earth System Research Laboratory (CO_2 data); GISS Surface Temperature Analysis, NASA (temperature data)

a Draw a scatter diagram to represent this data.

b Describe the correlation between c and t.

c Interpret your answer to part **b**.

(E) **3** The table below shows the packing times for a particular employee for a random sample of orders in a mail order company.

Number of items (n)	2	3	3	4	5	5	6	7	8	8	8	9	11	13
Time (t min)	11	14	16	16	19	21	23	25	24	27	28	30	35	42

A scatter diagram was drawn to represent the data.

a Describe the correlation between number of items packed and time taken. **(1 mark)**

The equation of the regression line of t on n is $t = 6.3 + 2.64n$.

b Give an interpretation of the value 2.64. **(1 mark)**

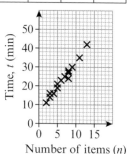

E **4** Energy consumption is claimed to be a good predictor of Gross National Product.
An economist recorded the energy consumption (x) and the Gross National Product (y) for eight countries. The data is shown in the table.

Energy consumption (x)	3.4	7.7	12.0	75	58	67	113	131
Gross National Product (y)	55	240	390	1100	1390	1330	1400	1900

The equation of the regression line of y on x is $y = 225 + 12.9x$.

The economist uses this regression equation to estimate the energy consumption of a country with a Gross National Product of 3500.

Give two reasons why this may not be a valid estimate. **(2 marks)**

E **5** The table shows average monthly temperature, t (°C), and the number of pairs of gloves, g, a shop sells each month.

t (°C)	6	6	50	10	13	16	18	19	16	12	9	7
g	81	58	50	42	19	21	4	2	20	33	58	65

The following statistics were calculated for the data on temperature:
mean = 15.2, standard deviation = 11.4
An outlier is an observation which lies ±2 standard deviations from the mean.

a Show that $t = 50$ is an outlier. **(1 mark)**

b Give a reason whether or not this outlier should be omitted from the data. **(1 mark)**

The equation of the regression line of t on g for the remaining data is $t = 18.4 - 0.18g$.

c Give an interpretation of the value -0.18 in this regression equation. **(1 mark)**

E **6** James placed different masses (m) on a spring and measured the resulting length of the spring (s) in centimetres. The smallest mass was 20 g and the largest mass was 100 g.

He found the equation of the regression line of s on m to be $s = 44 + 0.2m$.

a Interpret the values 44 and 0.2 in this context. **(2 marks)**

b Explain why it would not be sensible to use the regression equation to work out:
  i the value of s when $m = 150$ **ii** the value of m when $s = 60$. **(2 marks)**

E **7** A student is investigating the relationship between the price (y pence) of 100 g of chocolate and the percentage (x%) of cocoa solids in the chocolate.

The data obtained is shown in the table.

Chocolate brand	x (% cocoa)	y (pence)
A	10	35
B	20	55
C	30	40
D	35	100
E	40	60
F	50	90
G	60	110
H	70	130

a Draw a scatter diagram to represent this data. **(2 marks)**

The equation of the regression line of y on x is $y = 17.0 + 1.54x$.

b Draw the regression line on your diagram. **(2 marks)**

The student believes that one brand of chocolate is overpriced and uses the regression line to suggest a fair price for this brand.

c Suggest, with a reason, which brand is overpriced. **(1 mark)**

d Comment on the validity of the student's method for suggesting a fair price. **(1 mark)**

Large data set

You will need access to the large data set and spreadsheet software to answer these questions.

1 Investigate the relationship between daily mean windspeed, w, and daily maximum gust, g, in Leeming in 2015.
   a Draw a scatter diagram of w against g for the entire data set for Leeming in 2015.
   b Describe the correlation shown.
   c Comment on whether there is likely to be a causal relationship between mean windspeed and maximum gust.
   The equation of the regression line of g on w is given by $g = 4.97 + 2.15w$.
   d Use the equation of the regression line to predict the maximum gust on a day when the mean windspeed is:
      i 0.5 knots   ii 5 knots   iii 12 knots   iv 40 knots.
   e Comment on the accuracy of each prediction in part **d**.
   f Calculate the equation of the regression line of w on g, and use it to predict the mean windspeed on a day when the maximum gust was 30 knots.

> **Hint** You can use the SLOPE and INTERCEPT functions in some spreadsheets to find the values of a and b in a regression equation.

2 Use a similar approach to investigate the daily total sunshine and daily mean total cloud cover in Heathrow in 1987.
   a Use a regression model to suggest values for the missing total sunshine data in the first half of May.
   b Do you think there is a causal relationship between these two variables? Give a reason for your answer.

Summary of key points

1 **Bivariate data** is data which has pairs of values for two variables.

2 **Correlation** describes the nature of the linear relationship between two variables.

3 When two variables are correlated, you need to consider the context of the question and use your common sense to determine whether they have a causal relationship.

4 The **regression line** of y on x is written in the form $y = a + bx$.

5 The coefficient b tells you the change in y for each unit change in x.
   • If the data is positively correlated, b will be positive.
   • If the data is negatively correlated, b will be negative.

6 You should only use the regression line to make predictions for values of the dependent variable that are within the range of the given data.

Probability

Objectives

After completing this chapter you should be able to:

* Calculate probabilities for single events → **pages 70–72**
* Draw and interpret Venn diagrams → **pages 72–75**
* Understand mutually exclusive and independent events, and determine whether two events are independent → **pages 75–78**
* Use and understand tree diagrams → **pages 78–80**

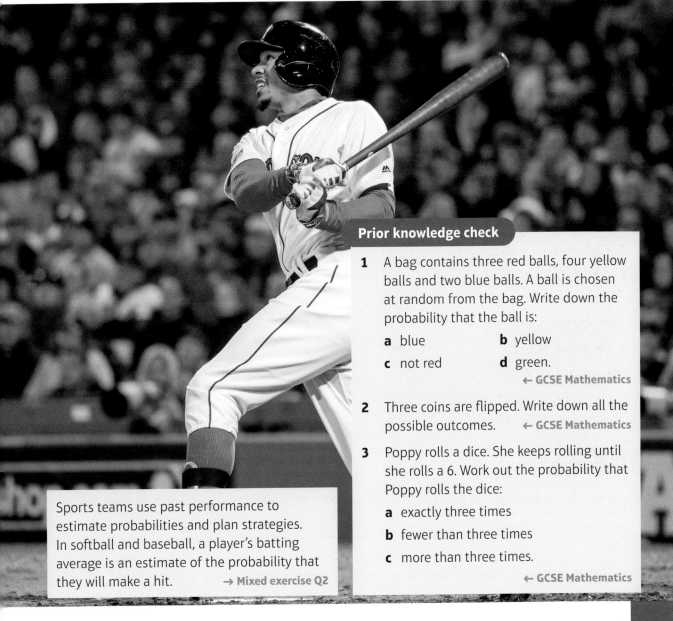

Prior knowledge check

1. A bag contains three red balls, four yellow balls and two blue balls. A ball is chosen at random from the bag. Write down the probability that the ball is:

   a blue   **b** yellow

   c not red   **d** green.
   ← **GCSE Mathematics**

2. Three coins are flipped. Write down all the possible outcomes. ← **GCSE Mathematics**

3. Poppy rolls a dice. She keeps rolling until she rolls a 6. Work out the probability that Poppy rolls the dice:

   a exactly three times

   b fewer than three times

   c more than three times.
   ← **GCSE Mathematics**

Sports teams use past performance to estimate probabilities and plan strategies. In softball and baseball, a player's batting average is an estimate of the probability that they will make a hit. → **Mixed exercise Q2**

5.1 Calculating probabilities

If you want to predict the chance of something happening, you use probability.

An **experiment** is a repeatable process that gives rise to a number of **outcomes**.

An **event** is a collection of one or more outcomes.

A **sample space** is the set of all possible outcomes.

Where outcomes are **equally likely** the probability of an event is the number of outcomes in the event divided by the total number of possible outcomes.

All events have probability between 0 (impossible) and 1 (certain). Probabilities are usually written as fractions or decimals.

Example 1

Two fair spinners each have four sectors numbered 1 to 4. The two spinners are spun together and the sum of the numbers indicated on each spinner is recorded.

Find the probability of the spinners indicating a sum of:

a exactly 5   **b** more than 5.

Draw a sample space diagram showing all possible outcomes.

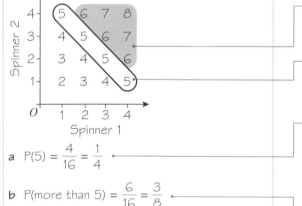

There are 4 × 4 = 16 points. Each of these points is equally likely as the spinners are fair.

There are 4 outcomes for part **a**.

a $P(5) = \dfrac{4}{16} = \dfrac{1}{4}$

There are four 5s and 16 outcomes altogether. P( ) is short for 'probability of'. The answer P(5) can also be written as 0.25.

b $P(\text{more than } 5) = \dfrac{6}{16} = \dfrac{3}{8}$

There are six sums more than 5 for this part (shaded blue). They form the top right corner of the diagram.

The answer can also be written as 0.375.

Example 2

The table shows the times taken, in minutes, for a group of students to complete a number puzzle.

Time, t (min)	$5 \leqslant t < 7$	$7 \leqslant t < 9$	$9 \leqslant t < 11$	$11 \leqslant t < 13$	$13 \leqslant t < 15$
Frequency	6	13	12	5	4

A student is chosen at random. Find the probability that they finished the number puzzle:

a in under 9 minutes   **b** in over 10.5 minutes.

a P(finished in under 9 minutes) = $\frac{19}{40}$

There are 40 students overall.

$6 + 13 = 19$ finished in under 9 minutes.

b $3 + 5 + 4 = 12$

P(finished in over 10.5 minutes) = $\frac{12}{40} = \frac{3}{10}$

Problem-solving

Use interpolation: 10.5 minutes lies $\frac{3}{4}$ of the way through the $9 \leqslant t < 11$ class, so $\frac{1}{4}$ of 12 is 3. Your answer is an estimate because you don't know the exact number of students who took longer than 10.5 seconds.

Exercise 5A

1 Two coins are tossed. Find the probability of both coins showing the same outcome.

2 Two six-sided dice are thrown and their product, X, is recorded.

   a Draw a sample space diagram showing all the possible outcomes of this experiment.

   b Find the probability of each event:

   i $X = 24$ ii $X < 5$ iii X is even.

(P) 3 The masses of 140 adult Bullmastiffs are recorded in a table.
One dog is chosen at random.

   a Find the probability that the dog has a mass of 54 kg or more.

   b Find the probability that the dog has a mass between 48 kg and 57 kg.

The probability that a Rottweiler chosen at random has a mass under 53 kg is 0.54.

   c Is it more or less likely that a Bullmastiff chosen at random has a mass under 53 kg? State one assumption that you have made in making your decision.

Mass, m (kg)	Frequency
$45 \leqslant m < 48$	17
$48 \leqslant m < 51$	25
$51 \leqslant m < 54$	42
$54 \leqslant m < 57$	33
$57 \leqslant m < 60$	21
$60 \leqslant m < 63$	2

Hint

Use interpolation.

(P) 4 The lengths, in cm, of 240 koalas are recorded in a table.

One koala is chosen at random.

   a Find the probability that the koala is female.

   b Find the probability that the koala is less than 80 cm long.

   c Find the probability that the koala is a male between 75 cm and 85 cm long.

Koalas under 72 cm long are called juvenile.

   d Estimate the probability that a koala chosen at random is juvenile. State one assumption you have made in making your estimate.

Length, l (cm)	Frequency (male)	Frequency (female)
$65 \leqslant l < 70$	4	14
$70 \leqslant l < 75$	20	15
$75 \leqslant l < 80$	24	32
$80 \leqslant l < 85$	47	27
$85 \leqslant l < 90$	31	26

E/P **5** The histogram shows the distribution of masses, in kg, of 70 adult cats.

  a Find the probability that a cat chosen at random has a mass more than 5 kg.
  (2 marks)

  b Estimate the probability that a cat chosen at random has a mass less than 6.5 kg. **(3 marks)**

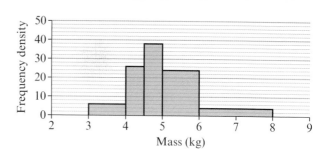

Challenge

Samira picks one card at random from group A and one card at random from group B.

She records the product, Y, of the two cards as the result of her experiment. Given that x is an integer and that P(Y is even) = P($Y \geqslant 20$), find the possible values of x.

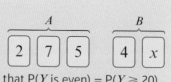

5.2 Venn diagrams

■ **A Venn diagram can be used to represent events graphically. Frequencies or probabilities can be placed in the regions of the Venn diagram.**

Venn diagrams are named after the English mathematician John Venn (1834–1923).

A rectangle represents the sample space, S, and it contains closed curves that represent events.

For events A and B in a sample space S:

1 The event A **and** B

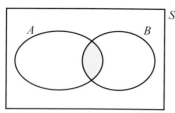

This event is also called the **intersection** of A and B. It represents the event that both A and B occur.

2 The event A **or** B

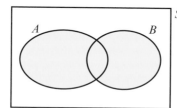

This event is also called the **union** of A and B.

It represents the event that either A or B, or both, occur.

3 The event **not** A

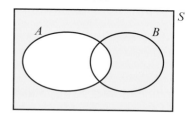

This event is also called the **complement** of A. It represents the event that A does not occur.

P(not A) = 1 − P(A)

You can write numbers of outcomes (frequencies) or the probability of the events in a Venn diagram to help solve problems.

Example 3

In a class of 30 students, 7 are in the choir, 5 are in the school band and 2 are in the choir and the band. A student is chosen at random from the class.

a Draw a Venn diagram to represent this information.

b Find the probability that:
  i the student is not in the band   **ii** the student is not in the choir or the band.

a

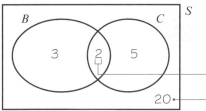

Put the number in both the choir and the band in the intersection of B and C.

This region represents the events in the sample space that are not in C or B:

$30 - (3 + 2 + 5) = 20$.

b i A student not in the band is not B.

$$P(\text{not } B) = \frac{25}{30} = \frac{5}{6}$$

There are $5 + 20 = 25$ outcomes not in B, out of 30 equally likely outcomes.

ii P(student is not in the choir or the band)

$$= \frac{20}{30} = \frac{2}{3}$$

20 outcomes are in neither event.

Example 4

A vet surveys 100 of her clients. She finds that:

25 own dogs   15 own dogs and cats
11 own dogs and fish   53 own cats
10 own cats and fish
7 own dogs, cats and fish
40 own fish

A client is chosen at random.
Find the probability that the client:
a owns dogs only
b does not own fish
c does not own dogs, cats or fish.

Problem-solving

You can use a Venn diagram with probabilities to solve this problem, but it could also be solved using the number of outcomes. There are 7 clients who own all three pets. Start with 0.07 in the intersection of all three events.

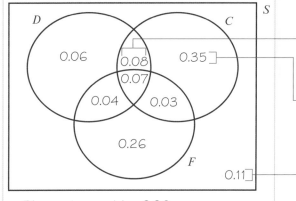

Work outwards to the intersections.
$0.15 - 0.07 = 0.08$

Each of 'dogs only', 'cats only' and 'fish only' can be worked out by further subtractions:
$0.53 - (0.08 + 0.07 + 0.03) = 0.35$ for 'cats only'

As the probability of the whole sample space is 1, the final area is $1 - (0.26 + 0.04 + 0.07 + 0.03 + 0.06 + 0.08 + 0.35) = 0.11$

a P(owns dogs only) = 0.06

b P(does not own fish) = $1 - 0.4 = 0.6$

c P(does not own dogs, cats or fish) = 0.11

This is the value on the Venn diagram outside D, C and F.

1 There are 25 students in a certain tutor group at Philips College. There are 16 students in the tutor group studying German, 14 studying French and 6 students studying both French and German.

   a Draw a Venn diagram to represent this information.

   b Find the probability that a randomly chosen student in the tutor group:

     i studies French   **ii** studies French and German

     iii studies French but not German   **iv** does not study French or German.

2 There are 125 diners in a restaurant who were surveyed to find out if they had ordered garlic bread, beer or cheesecake:

        15 diners had ordered all three items   20 had ordered beer and cheesecake
        43 diners had ordered garlic bread   26 had ordered garlic bread and cheesecake
        40 diners had ordered beer   25 had ordered garlic bread and beer
        44 diners had ordered cheesecake

   a Draw a Venn diagram to represent this information.

   A diner is chosen at random. Find the probability that the diner ordered:

   b **i** all three items   **ii** beer but not cheesecake and not garlic bread

     iii garlic bread and beer but not cheesecake   **iv** none of these items.

3 A group of 275 people at a music festival were asked if they play guitar, piano or drums:

        one person plays all three instruments   15 people play piano only
        65 people play guitar and piano   20 people play guitar only
        10 people play piano and drums   35 people play drums only
        30 people play guitar and drums

   a Draw a Venn diagram to represent this information.

   b A festival goer is chosen at random from the group.
     Find the probability that the person chosen:

     i plays the piano   **ii** plays at least two of guitar, piano and drums

     iii plays exactly one of the instruments   **iv** plays none of the instruments.

(P) 4 The probability that a child in a school has blue eyes is 0.27 and the probability that they have blonde hair is 0.35. The probability that the child will have blonde hair or blue eyes or both is 0.45. A child is chosen at random from the school. Find the probability that the child has:

   a blonde hair and blue eyes

   b blonde hair but not blue eyes

   c neither feature.

> **Hint** Draw a Venn diagram to help you.

(E/P) 5 A patient going in to a doctor's waiting room reads *Hiya* magazine with probability 0.6 and *Dakor* magazine with probability 0.4. The probability that the patient reads either one or both of the magazines is 0.7. Find the probability that the patient reads:

   a both magazines   **(2 marks)**

   b *Hiya* magazine only.   **(2 marks)**

E/P **6** The Venn diagram shows the probabilities of members of a sports club taking part in various activities.

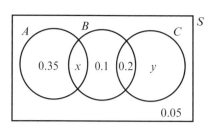

A represents the event that the member takes part in archery.

B represents the event that the member takes part in badminton.

C represents the event that the member takes part in croquet.

Given that P(*B*) = 0.45:

a find *x* **(1 mark)**

b find *y*. **(2 marks)**

E/P **7** The Venn diagram shows the probabilities that students at a sixth-form college study certain subjects.

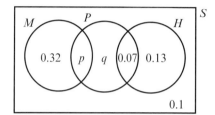

M represents the event that the student studies mathematics.

P represents the event that the student studies physics.

H represents the event that the student studies history.

Given that P(*M*) = P(*P*), find the values of *p* and *q*. **(4 marks)**

Challenge

The Venn diagram shows the probabilities of a group of children liking three types of sweet.

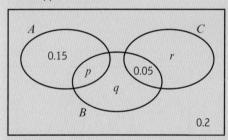

Given that P(*B*) = 2P(*A*) and that P(not *C*) = 0.83, find the values of *p*, *q* and *r*.

5.3 Mutually exclusive and independent events

When events have no outcomes in common they are called **mutually exclusive**.

In a Venn diagram, the closed curves do not overlap and you can use a simple addition rule to work out combined probabilities:

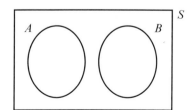

- **For mutually exclusive events, P(*A* or *B*) = P(*A*) + P(*B*).**

When one event has no effect on another, they are **independent**. Therefore if *A* and *B* are independent, the probability of *A* happening is the same whether or not *B* happens.

- **For independent events, P(*A* and *B*) = P(*A*) × P(*B*).**

You can use this **multiplication rule** to determine whether events are independent.

Example 5

Events A and B are mutually exclusive and $P(A) = 0.2$ and $P(B) = 0.4$.

Find: **a** P(A or B)   **b** P(A but not B)   **c** P(neither A nor B)

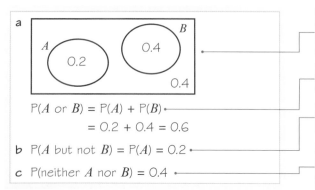

a

P(A or B) = P(A) + P(B)

= 0.2 + 0.4 = 0.6

b P(A but not B) = P(A) = 0.2

c P(neither A nor B) = 0.4

A and B are mutually exclusive so the closed curves do not intersect.

Use the simple addition rule.

Everything in A is 'not B'.

This is everything outside of both circles: $1 - P(A \text{ or } B)$.

Example 6

Events A and B are independent and $P(A) = \frac{1}{3}$ and $P(B) = \frac{1}{5}$.

Find P(A and B).

$$P(A \text{ and } B) = P(A) \times P(B) = \frac{1}{3} \times \frac{1}{5} = \frac{1}{15}$$

A and B are independent so you can use the multiplication rule for independent events.

Example 7

The Venn diagram shows the number of students in a particular class who watch any of three popular TV programmes.

a Find the probability that a student chosen at random watches B or C or both.

b Determine whether watching A and watching B are statistically independent.

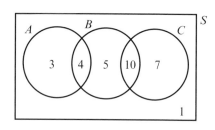

a $4 + 5 + 10 + 7 = 26$

P(watches B or C or both) $= \frac{26}{30} = \frac{13}{15}$

b $P(A) = \frac{3 + 4}{30} = \frac{7}{30}$

$P(B) = \frac{4 + 5 + 10}{30} = \frac{19}{30}$

$P(A \text{ and } B) = \frac{4}{30} = \frac{2}{15}$

$P(A) \times P(B) = \frac{7}{30} \times \frac{19}{30} = \frac{133}{900}$

So P(A and B) $\neq$ P(A) $\times$ P(B)

Therefore watching A and watching B are *not* independent.

Take the probabilities from the Venn diagram.

Multiply the two probabilities and check whether they give the same answer as P(A and B).

Problem-solving

Show your calculations and then write down a conclusion stating whether or not the events are independent.

Exercise 5C

1 Events A and B are mutually exclusive. $P(A) = 0.2$ and $P(B) = 0.5$.
   a Draw a Venn diagram to represent these two events.
   b Find $P(A \text{ or } B)$.
   c Find $P(\text{neither } A \text{ nor } B)$.

2 Two fair dice are rolled and the result on each die is recorded. Show that the events 'the sum of the scores on the dice is 4' and 'both dice land on the same number' are not mutually exclusive.

3 $P(A) = 0.5$ and $P(B) = 0.3$. Given that events A and B are independent, find $P(A \text{ and } B)$.

4 $P(A) = 0.15$ and $P(A \text{ and } B) = 0.045$. Given that events A and B are independent, find $P(B)$.

5 The Venn diagram shows the number of children in a play group that like playing with bricks (B), action figures (F) or trains (T).
   a State, with a reason, which two types of toy are mutually exclusive.
   b Determine whether the events 'plays with bricks' and 'plays with action figures' are independent.

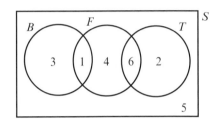

6 The Venn diagram shows the probabilities that a group of students like pasta (A) or pizza (B).
   a Write down the value of x. **(1 mark)**
   b Determine whether the events 'like pasta' and 'like pizza' are independent. **(3 marks)**

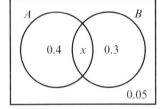

7 S and T are two events such that $P(S) = 0.3$, $P(T) = 0.4$ and $P(S \text{ but not } T) = 0.18$.
   a Show that S and T are independent.
   b Find:
     i $P(S \text{ and } T)$ ii $P(\text{neither } S \text{ nor } T)$.

8 W and X are two events such that $P(W) = 0.5$, $P(W \text{ and not } X) = 0.25$ and $P(\text{neither } W \text{ nor } X) = 0.3$. State, with a reason, whether W and X are independent events.
   (3 marks)

9 The Venn diagram shows the probabilities of members of a social club taking part in charitable activities.

   A represents taking part in an archery competition.

   R represents taking part in a raffle.

   F represents taking part in a fun run.

   The probability that a member takes part in the archery competition or the raffle is 0.6.
   a Find the value of x and the value of y. **(2 marks)**
   b Show that events R and F are not independent. **(3 marks)**

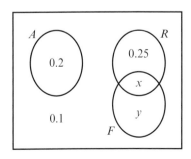

(P) 10 Given that A and B are independent, find the two possible values for p and q.

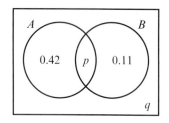

Challenge

A and B are independent events in a sample space S. Given that A and B are independent, prove that:

a A and 'not B' are independent

b 'not A' and 'not B' are independent.

5.4 Tree diagrams

- **A tree diagram can be used to show the outcomes of two (or more) events happening in succession.**

Example 8

A bag contains seven green beads and five blue beads. A bead is taken from the bag at random and not replaced. A second bead is then taken from the bag.

Find the probability that:

a both beads are green

b the beads are different colours.

a

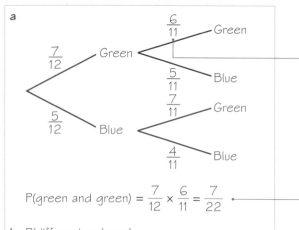

Draw a tree diagram to show the events.

There are now only 6 green beads and 11 beads in total.

$$P(\text{green and green}) = \frac{7}{12} \times \frac{6}{11} = \frac{7}{22}$$

Multiply along the branch of the tree diagram.

b P(different colours)

= P(green then blue) + P(blue then green)

$$= \frac{7}{12} \times \frac{5}{11} + \frac{5}{12} \times \frac{7}{11} = \frac{35}{66}$$

Multiply along each branch and add the two probabilities.

Exercise 5D

1 A bag contains three red beads and five blue beads. A bead is chosen at random from the bag, the colour is recorded and the bead is replaced. A second bead is chosen and the colour recorded.

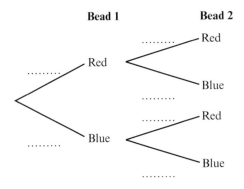

Bead 1   **Bead 2**

a Copy and complete this tree diagram to show the outcomes of the experiment.

b Find the probability that both beads are blue.

c Find the probability that the second bead is blue.

2 A box contains nine cards numbered 1 to 9. A card is drawn at random and not replaced. It is noted whether the number is odd or even. A second card is drawn and it is also noted whether this number is odd or even.

a Draw a tree diagram to represent this experiment.

Hint The first card is not replaced.

b Find the probability that both cards are even.

c Find the probability that one card is odd and the other card is even.

3 The probability that Charlie takes the bus to school is 0.4. If he doesn't take the bus, he walks. The probability that Charlie is late to school if he takes the bus is 0.2. The probability he is late to school if he walks is 0.3.

a Draw a tree diagram to represent this information.

b Find the probability that Charlie is late to school.

(E) 4 Mr Dixon plays golf. The probability that he scores par or under on the first hole is 0.7. If he scores par or under on the first hole, the probability he scores par or under on the second hole is 0.8. If he doesn't score par or under on the first hole, the probability that he scores par or under on the second hole is 0.4.

a Draw a tree diagram to represent this information. **(3 marks)**

b State whether the events 'scores par or under on the first hole' and 'scores par or under par on the second hole' are independent. **(1 mark)**

c Find the probability that Mr Dixon scores par or under on only one hole. **(3 marks)**

(E/P) 5 A biased coin is tossed three times and it is recorded whether it falls heads or tails. P(heads) = $\frac{1}{3}$

a Draw a tree diagram to represent this experiment. **(3 marks)**

b Find the probability that the coin lands on heads all three times. **(1 mark)**

c Find the probability that the coin lands on heads only once. **(2 marks)**

The whole experiment is repeated for a second trial.

d Find the probability of obtaining either 3 heads or 3 tails in both trials. **(3 marks)**

(E/P) **6** A bag contains 13 tokens, 4 coloured blue, 3 coloured red and 6 coloured yellow. Two tokens are drawn from the bag without replacement.

    a Find the probability that both tokens are yellow. **(2 marks)**

A third token is drawn from the bag.

    b Write down the probability that the third token is yellow, given that the first two are yellow. **(1 mark)**

    c Find the probability that all three tokens are different colours. **(4 marks)**

Mixed exercise 5

(E/P) **1** There are 15 coloured beads in a bag; seven beads are red, three are blue and five are green. Three beads are selected at random from the bag and replaced. Find the probability that:

    a the first and second beads chosen are red and the third bead is blue or green **(3 marks)**

    b one red, one blue and one green bead are chosen. **(3 marks)**

2 A baseball player has a batting average of 0.341. This means her probability of making a hit when she bats is 0.341. She bats three times in one game. Estimate the probability that:

    a she makes three hits

    b she makes no hits

    c she makes at least one hit.

(P) **3** The scores of 250 students in a test are recorded in a table.

One student is chosen at random.

    a Find the probability that the student is female.

    b Find the probability that the student scored less than 35.

    c Find the probability that the student is a male and scored between 25 and 34.

Score, s	Frequency (male)	Frequency (female)
$20 \leqslant s < 25$	7	8
$25 \leqslant s < 30$	15	13
$30 \leqslant s < 35$	18	19
$35 \leqslant s < 40$	25	30
$40 \leqslant s < 45$	30	26
$45 \leqslant s < 50$	27	32

In order to pass the test, students must score 37 or more.

    d Estimate the probability that a student chosen at random passes the test. State one assumption you have made in making your estimate.

(E/P) **4** The histogram shows the distribution of masses, in kg, of 50 newborn babies.

    a Find the probability that a baby chosen at random has a mass greater than 3 kg. **(2 marks)**

    b Estimate the probability that a baby chosen at random has a mass less than 3.75 kg. **(3 marks)**

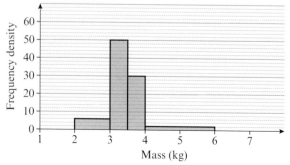

(E) **5** A study was made of a group of 150 children to determine which of three cartoons they watch on television. The following results were obtained:

35 watch Toontime   14 watch Porky and Skellingtons

54 watch Porky   12 watch Toontime and Skellingtons

62 watch Skellingtons   4 watch Toontime, Porky and Skellingtons

9 watch Toontime and Porky

  a Draw a Venn diagram to represent this data.   **(4 marks)**

  b Find the probability that a randomly selected child from the study watches:

    i none of the three cartoons   **(2 marks)**

    ii no more than one of the cartoons.   **(2 marks)**

(P) **6** The events A and B are such that $P(A) = \frac{1}{3}$ and $P(B) = \frac{1}{4}$. $P(A \text{ or } B \text{ or both}) = \frac{1}{2}$.

  a Represent these probabilities on a Venn diagram.

  b Show that A and B are independent.

(E) **7** The Venn diagram shows the number of students who like either cricket (C), football (F) or swimming (S).

  a Which two sports are mutually exclusive?   **(1 mark)**

  b Determine whether the events 'likes cricket' and 'likes football' are independent.   **(3 marks)**

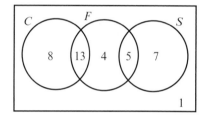

(E/P) **8** For events J and K, $P(J \text{ or } K \text{ or both}) = 0.5$, $P(K \text{ but not } J) = 0.2$ and $P(J \text{ but not } K) = 0.25$.

  a Draw a Venn diagram to represent events J and K and the sample space S.   **(3 marks)**

  b Determine whether events J and K are independent.   **(3 marks)**

(E) **9** A survey of a group of students revealed that 85% have a mobile phone, 60% have an MP3 player and 5% have neither phone nor MP3 player.

  a Find the proportion of students who have both gadgets.   **(2 marks)**

  b Draw a Venn diagram to represent this information.   **(3 marks)**

  c A student is chosen at random. Find the probability that they only own a mobile phone.   **(2 marks)**

  d Are the events 'own a mobile phone' and 'own an MP3 player' independent? Justify your answer.   **(3 marks)**

(E/P) **10** The Venn diagram shows the probabilities that a group of children like cake (A) or crisps (B).

Determine whether the events 'like cake' and 'like crisps' are independent.   **(3 marks)**

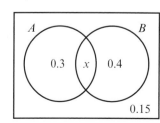

(E/P) **11** A computer game has three levels and one of the objectives of every level is to collect a diamond. The probability that Becca collects a diamond on the first level is $\frac{4}{5}$, the second level is $\frac{2}{3}$ and the third level is $\frac{1}{2}$. The events are independent.

    a Draw a tree diagram to represent Becca collecting diamonds on the three levels of the game. **(4 marks)**

    b Find the probability that Becca:

       i collects all three diamonds **(2 marks)**

       ii collects only one diamond. **(3 marks)**

    c Find the probability that she collects at least two diamonds each time she plays. **(3 marks)**

(P) **12** In a factory, machines A, B and C produce electronic components. Machine A produces 16% of the components, machine B produces 50% of the components and machine C produces the rest. Some of the components are defective. Machine A produces 4%, machine B 3% and machine C 7% defective components.

    a Draw a tree diagram to represent this information.

    b Find the probability that a randomly selected component is:

       i produced by machine B and is defective   **ii** defective.

Challenge

The members of a cycling club are married couples. For any married couple in the club, the probability that the husband is retired is 0.7 and the probability that the wife is retired 0.4. Given that the wife is retired, the probability that the husband is retired is 0.8.

Two married couples are chosen at random.

Find the probability that only one of the two husbands and only one of the two wives is retired.

Summary of key points

1 A **Venn diagram** can be used to represent events graphically. Frequencies or probabilities can be placed in the regions of the Venn diagram.

2 For **mutually exclusive** events, P(A or B) = P(A) + P(B).

3 For **independent** events, P(A and B) = P(A) × P(B).

4 A **tree diagram** can be used to show the outcomes of two (or more) events happening in succession.

Statistical distributions

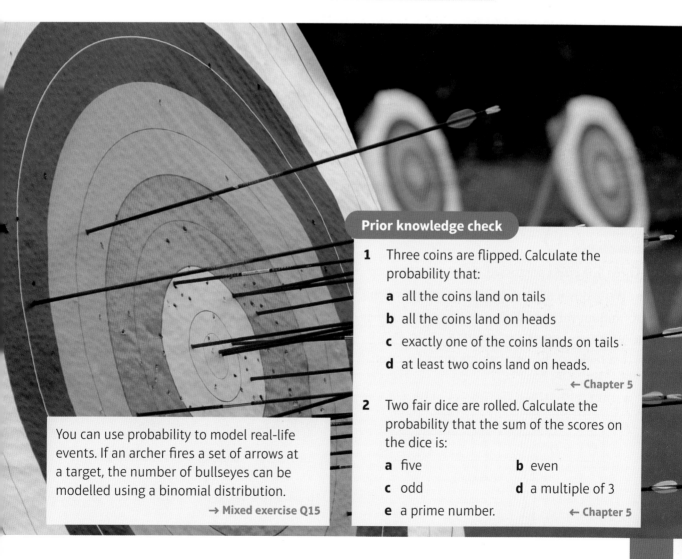

6

→ pages 84–88

→ page 88

→ pages 89–91

→ pages 91–94

Objectives

After completing this chapter you should be able to:

* Understand and use simple discrete probability distributions including the discrete uniform distribution
* Understand the binomial distribution as a model and comment on appropriateness
* Calculate individual probabilities for the binomial distribution
* Calculate cumulative probabilities for the binomial distribution

Prior knowledge check

1. Three coins are flipped. Calculate the probability that:
   a all the coins land on tails
   b all the coins land on heads
   c exactly one of the coins lands on tails
   d at least two coins land on heads.
   ← Chapter 5

2. Two fair dice are rolled. Calculate the probability that the sum of the scores on the dice is:
   a five   b even
   c odd   d a multiple of 3
   e a prime number.   ← Chapter 5

You can use probability to model real-life events. If an archer fires a set of arrows at a target, the number of bullseyes can be modelled using a binomial distribution.
→ Mixed exercise Q15

6.1 Probability distributions

A **random variable** is a variable whose value depends on the outcome of a random event.

- The range of values that a random variable can take is called its **sample space**.
- A **variable** can take any of a range of specific values.
- The variable is **discrete** if it can only take *certain* numerical values.
- The variable is **random** if the outcome is not known until the experiment is carried out.

> **Notation** Random variables are written using upper case letters, for example X or Y.
>
> The particular values the random variable can take are written using equivalent lower case letters, for example x or y.
>
> The probability that the random variable X takes a particular value x is written as $P(X = x)$.

- **A probability distribution fully describes the probability of any outcome in the sample space.**

The probability distribution for a discrete random variable can be described in a number of different ways. For example, take the random variable X = 'score when a fair dice is rolled'. It can be described:

- as a **probability mass function**: $P(X = x) = \frac{1}{6}$, $x = 1, 2, 3, 4, 5, 6$

- using a table:

x	1	2	3	4	5	6
$P(X = x)$	$\frac{1}{6}$	$\frac{1}{6}$	$\frac{1}{6}$	$\frac{1}{6}$	$\frac{1}{6}$	$\frac{1}{6}$

- using a diagram:

All of these representations show the probability that the random variable takes any given value in its sample space.

When all of the probabilities are the same, as in this example, the distribution is known as a **discrete uniform distribution**.

Example 1

Three fair coins are tossed.

a Write down all the possible outcomes when the three coins are tossed.

A random variable, X, is defined as the number of heads when the three coins are tossed.

b Write the probability distribution of X as:

   i a table   **ii** a probability mass function.

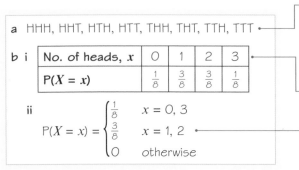

a HHH, HHT, HTH, HTT, THH, THT, TTH, TTT

These are the outcomes of the experiment.

b i

No. of heads, x	0	1	2	3
$P(X = x)$	$\frac{1}{8}$	$\frac{3}{8}$	$\frac{3}{8}$	$\frac{1}{8}$

X is the number of heads, so the sample space of X is {0, 1, 2, 3}.

These are the values of the random variable.

ii

$$P(X = x) = \begin{cases} \frac{1}{8} & x = 0, 3 \\ \frac{3}{8} & x = 1, 2 \\ 0 & otherwise \end{cases}$$

These are the values the random variable can take.

- **The sum of the probabilities of all outcomes of an event add up to 1. For a random variable X, you can write $\sum P(X = x) = 1$ for all x.**

Example 2

A biased four-sided dice with faces numbered 1, 2, 3 and 4 is rolled. The number on the bottom-most face is modelled as a random variable X.

Given that $P(X = x) = \dfrac{k}{x}$:

a Find the value of k.

b Give the probability distribution of X in table form.

c Find the probability that:

   i $X > 2$ **ii** $1 < X < 4$ **iii** $X > 4$

---

a The probability distribution will be:

x	1	2	3	4
$P(X = x)$	$\dfrac{k}{1}$	$\dfrac{k}{2}$	$\dfrac{k}{3}$	$\dfrac{k}{4}$

$\dfrac{k}{1} + \dfrac{k}{2} + \dfrac{k}{3} + \dfrac{k}{4} = 1$

$k\left(1 + \dfrac{1}{2} + \dfrac{1}{3} + \dfrac{1}{4}\right) = 1$

$k\left(\dfrac{12 + 6 + 4 + 3}{12}\right) = 1$

$k = \dfrac{12}{25}$

> Since this is a probability distribution,
> $\sum P(X = x) = 1$

b The probability distribution is:

x	1	2	3	4
$P(X = x)$	$\dfrac{12}{25}$	$\dfrac{6}{25}$	$\dfrac{4}{25}$	$\dfrac{3}{25}$

> **Problem-solving**
>
> Write an equation and solve it to find the value of k. Then substitute this value of k into
> $P(X = x) = \dfrac{k}{x}$ for each x to find the probabilities.

c **i** $X > 2$ is the same as getting 3 or 4 so

$P(X > 2) = \dfrac{4}{25} + \dfrac{3}{25} = \dfrac{7}{25}$

> Consider all the values of x that satisfy this condition. Add the probabilities to find $P(X > 2)$.

ii $1 < X < 4$ is the same as getting 2 or 3

$P(1 < X < 4) = \dfrac{6}{25} + \dfrac{4}{25} = \dfrac{10}{25} = \dfrac{2}{5}$

iii There are no elements in the sample space that satisfy $X > 4$ so

$P(X > 4) = 0$

> **Watch out** This random variable only **models** the behaviour of the dice. The outcomes from experiments in real life will never exactly fit the model, but the model provides a useful way of analysing possible outcomes.

Example 3

This spinner is spun until it lands on red or has been spun four times in total.

Find the probability distribution of the random variable S, the number of times the spinner is spun.

Problem-solving

Read the definition of the random variable carefully. Here it is the number of spins.

$P(S = 1)$ is the probability that the spinner lands on red the first time:

$P(S = 1) = \dfrac{2}{5}$

On any given spin, $P(\text{Red}) = \dfrac{2}{5}$ and $P(\text{Blue}) = \dfrac{3}{5}$.

If the spinner lands on red on the second spin it must land on blue on the first spin:

$P(S = 2) = \dfrac{3}{5} \times \dfrac{2}{5} = \dfrac{6}{25}$

Each spin is an independent event so $P(\text{Blue then red}) = P(\text{Blue}) \times P(\text{Red})$

Likewise for landing on red on the third spin:

$P(S = 3) = \dfrac{3}{5} \times \dfrac{3}{5} \times \dfrac{2}{5} = \dfrac{18}{125}$

B, B, R is the only outcome for which $S = 3$.

The experiment stops after 4 spins so:

$P(S = 4) = 1 - \left(\dfrac{2}{5} + \dfrac{6}{25} + \dfrac{18}{125} \right) = \dfrac{27}{125}$

The sample space of S is $\{1, 2, 3, 4\}$.
So $P(S = 4) = 1 - P(S = 1, 2 \text{ or } 3)$.

x	1	2	3	4
$P(S = s)$	$\dfrac{2}{5}$	$\dfrac{6}{25}$	$\dfrac{18}{125}$	$\dfrac{27}{125}$

You have found $P(S = s)$ for all values in the sample space, so you have found the complete probability distribution. You can summarise it in a table.

Exercise 6A

1 Write down whether or not each of the following is a discrete random variable. Give a reason for your answer.

  a The height, X cm, of a seedling chosen randomly from a group of plants.

  b The number of times, R, a six is rolled when a fair dice is rolled 100 times.

  c The number of days, W, in a given week.

2 A fair dice is thrown four times and the number of times it falls with a 6 on the top, Y, is noted. Write down the sample space of Y.

3 A bag contains two discs with the number 2 on them and two discs with the number 3 on them. A disc is drawn at random from the bag and the number noted. The disc is returned to the bag. A second disc is then drawn from the bag and the number noted.

  a Write down all the possible outcomes of this experiment.

The discrete random variable X is defined as the sum of the two numbers.

  b Write down the probability distribution of X as:

    i a table   **ii** a probability mass function.

4 A discrete random variable X has the probability distribution shown in the table. Find the value of k.

x	1	2	3	4
$P(X = x)$	$\frac{1}{3}$	$\frac{1}{3}$	k	$\frac{1}{4}$

(E/P) 5 The random variable X has a probability function

$$P(X = x) = kx \qquad x = 1, 2, 3, 4.$$

Show that $k = \frac{1}{10}$.　　　　　　　　　　　　　　**(2 marks)**

(E/P) 6 The random variable X has a probability function

$$P(X = x) = \begin{cases} kx & x = 1, 3 \\ k(x - 1) & x = 2, 4 \end{cases}$$

where k is a constant.

a Find the value of k.　　　　　　　　　　　　　　**(2 marks)**

b Find $P(X > 1)$.　　　　　　　　　　　　　　**(2 marks)**

(P) 7 The discrete random variable X has a probability function

$$P(X = x) = \begin{cases} 0.1 & x = -2, -1 \\ \beta & x = 0, 1 \\ 0.2 & x = 2 \end{cases}$$

a Find the value of β.

b Construct a table giving the probability distribution of X.

c Find $P(-1 \leqslant X < 2)$.

(P) 8 A discrete random variable has a probability distribution shown in the table. Find the value of a.

x	0	1	2
$P(X = x)$	$\frac{1}{4} - a$	a	$\frac{1}{2} + a$

(P) 9 The random variable X can take any integer value from 1 to 50. Given that X has a discrete uniform distribution, find:

a $P(X = 1)$

b $P(X \geqslant 28)$

c $P(13 < X < 42)$

(E) 10 A discrete random variable X has the probability distribution shown in this table. Find:

x	0	1	2	3
$P(X = x)$	$\frac{1}{8}$	$\frac{1}{4}$	$\frac{1}{2}$	$\frac{1}{8}$

a $P(1 < X \leqslant 3)$　　　　　　　　　　　　　　**(1 mark)**

b $P(X < 2)$　　　　　　　　　　　　　　**(1 mark)**

c $P(X > 3)$　　　　　　　　　　　　　　**(1 mark)**

E/P **11** A biased coin is tossed until a head appears or it is tossed four times.

If $P(\text{Head}) = \dfrac{2}{3}$:

   a Write down the probability distribution of S, the number of tosses, in table form. **(4 marks)**

   b Find $P(S > 2)$. **(1 mark)**

P **12** A fair five-sided spinner is spun.

Given that the spinner is spun five times, write down, in table form, the probability distributions of the following random variables:

   a X, the number of times red appears

   b Y, the number of times yellow appears.

The spinner is now spun until it lands on blue, or until it has been spun five times. The random variable Z is defined as the number of spins in this experiment.

   c Find the probability distribution of Z.

E/P **13** Marie says that a random variable X has a probability distribution defined by the following probability mass function:

$$P(X = x) = \frac{2}{x^2}, \quad x = 2, 3, 4$$

   a Explain how you know that Marie's function does not describe a probability
      distribution. **(2 marks)**

   b Given that the correct probability mass function is in the form $P(X = x) = \dfrac{k}{x^2}, \quad x = 2, 3, 4$

      where k is a constant, find the exact value of k. **(2 marks)**

Challenge

The independent random variables X and Y have probability distributions

$$P(X = x) = \frac{1}{8}, \quad x = 1, 2, 3, 4, 5, 6, 7, 8 \qquad P(Y = y) = \frac{1}{y}, \quad y = 2, 3, 6$$

Find $P(X > Y)$.

> **Hint** X and Y are independent so the value taken by one does not affect the probabilities for the other.

6.2 The binomial distribution

When you are carrying out a number of trials in an experiment or survey, you can define a random variable X to represent the number of **successful trials**.

- **You can model X with a binomial distribution, B(n, p), if:**
  - **there are a fixed number of trials, n**
  - **there are two possible outcomes (success and failure)**
  - **there is a fixed probability of success, p**
  - **the trials are independent of each other**

> **Hint** If $P(\text{Success}) = p$ and there are only two outcomes then $P(\text{Failure}) = 1 - p$.

- If a random variable X has the binomial distribution $B(n, p)$ then its probability mass function is given by

$$P(X = r) = \binom{n}{r}p^r(1 - p)^{n-r}$$

n is sometimes called the **index** and p is sometimes called the **parameter**.

You can use your calculator to work out binomial probabilities. You can either use the rule given above, together with the ^nC_r function, or use the binomial probability distribution function directly.

Notation You write $X \sim B(n, p)$

Links $\binom{n}{r} = \dfrac{n!}{r!(n-r)!}$

It is sometimes written as nCr or ^nC_r. It represents the number of ways of selecting r successful outcomes from n trials.

← **Pure Year 1, Chapter 8**

Example 4

The random variable $X \sim B\left(12, \frac{1}{6}\right)$. Find:

a $P(X = 2)$ **b** $P(X = 9)$ **c** $P(X \leqslant 1)$

a $P(X = 2) = \binom{12}{2}\left(\frac{1}{6}\right)^2\left(\frac{5}{6}\right)^{10} = \dfrac{12!}{2!10!}\left(\frac{1}{6}\right)^2\left(\frac{5}{6}\right)^{10}$

         $= 0.296\,09\ldots$

         $= 0.296$ (3 s.f.)

Use the formula with $n = 12$, $p = \frac{1}{6}$ and $x = 2$.

b $P(X = 9) = \binom{12}{9}\left(\frac{1}{6}\right)^9\left(\frac{5}{6}\right)^3$

         $= 0.000\,012\,63\ldots$

         $= 0.000\,012\,6$ (3 s.f.)

Use the formula with $n = 12$, $p = \frac{1}{6}$ and $x = 9$.

c $P(X \leqslant 1) = P(X = 0) + P(X = 1)$

         $= \left(\frac{5}{6}\right)^{12} + \binom{12}{1}\left(\frac{1}{6}\right)^1\left(\frac{5}{6}\right)^{11}$

         $= 0.112\,156\ldots + 0.269\,17\ldots$

         $= 0.381\,33\ldots$

         $= 0.381$ (3 s.f.)

A binomial distribution can take any value from 0 up to n inclusive. So there are two possible outcomes that satisfy the inequality: $X = 0$ and $X = 1$.

Online Use the ^nC_r function on your calculator to work out binomial probabilities.

Example 5

The probability that a randomly chosen member of a reading group is left-handed is 0.15. A random sample of 20 members of the group is taken.

a Suggest a suitable model for the random variable X, the number of members in the sample who are left-handed. Justify your choice.

b Use your model to calculate the probability that:

  i exactly 7 of the members in the sample are left-handed

  ii fewer than two of the members in the sample are left-handed.

a The random variable can take two values, left-handed or right-handed.
There are a fixed number of trials, 20, and a fixed probability of success: 0.15.
Assuming each member in the sample is independent, a suitable model is
$X \sim B(20, 0.15)$

A binomial model is a suitable choice. State the assumptions that are necessary for the binomial model, and make sure that you specify the values of n and p.

b i $P(X = 7) = \binom{20}{7} \times (0.15)^7 (0.85)^{13}$

$= 0.016\,01...$

$= 0.0160$ (3 s.f.)

Online Work this out directly using the binomial probability distribution function on your calculator and entering $x = 7$, $n = 20$ and $p = 0.15$.

ii $P(X < 2) = P(X = 0) + P(X = 1)$

$= 0.038\,75... + 0.136\,79...$

$= 0.176$ (3 s.f.)

In this situation, 'fewer than two' means 0 or 1.

Exercise 6B

1 The random variable $X \sim B\left(8, \frac{1}{3}\right)$. Find:

  a $P(X = 2)$ **b** $P(X = 5)$ **c** $P(X \leqslant 1)$

2 The random variable $T \sim B\left(15, \frac{2}{3}\right)$. Find:

  a $P(T = 5)$ **b** $P(T = 10)$ **c** $P(3 \leqslant T \leqslant 4)$

3 A student suggests using a binomial distribution to model the following situations. Give a description of the random variable, state any assumptions that must be made and give possible values for n and p.

  a A sample of 20 bolts from a large batch is checked for defects. The production process should produce 1% of defective bolts.

  b Some traffic lights have three phases: stop 48% of the time, wait or get ready 4% of the time, and go 48% of the time. Assuming that you only cross a traffic light when it is in the go position, model the number of times that you have to wait or stop on a journey passing through 6 sets of traffic lights.

  c When Stephanie plays tennis with Timothy, on average one in eight of her serves is an 'ace'. How many 'aces' does Stephanie serve in the next 30 serves against Timothy?

4 State which of the following can be modelled with a binomial distribution and which cannot. Give reasons for your answers.

  a Given that 15% of people have blood that is Rhesus negative (Rh⁻), model the number of pupils in a statistics class of 14 who are Rh⁻.

  b You are given a fair coin and told to keep tossing it until you obtain 4 heads in succession. Model the number of tosses you need.

  c A certain car manufacturer produces 12% of new cars in the colour red, 8% in blue, 15% in white and the rest in other colours. You make a note of the colour of the first 15 new cars of this make. Model the number of red cars you observe.

(P) **5** A balloon manufacturer claims that 95% of his balloons will not burst when blown up. If you have 20 of these balloons to blow up for a birthday party:

   a What is the probability that none of them burst when blown up?

   b Find the probability that exactly 2 balloons burst.

(E/P) **6** The probability of a switch being faulty is 0.08. A random sample of 10 switches is taken from the production line.

   a Define a suitable distribution to model the number of faulty switches in this sample, and justify your choice. **(2 marks)**

   b Find the probability that the sample contains 4 faulty switches. **(2 marks)**

(E/P) **7** A particular genetic marker is present in 4% of the population.

   a State any assumptions that are required to model the number of people with this genetic marker in a sample of size n as a binomial distribution. **(2 marks)**

   b Using this model, find the probability of exactly 6 people having this marker in a sample of size 50. **(2 marks)**

(E/P) **8** A dice is biased so that the probability of it landing on a six is 0.3. Hannah rolls the dice 15 times.

   a State any assumptions that are required to model the number of sixes as a binomial distribution. State the distribution. **(2 marks)**

   b Find the probability that Hannah rolls exactly 4 sixes. **(2 marks)**

   c Find the probability that she rolls two or fewer sixes. **(3 marks)**

6.3 Cumulative probabilities

A **cumulative probability function** for a random variable X tells you the sum of all the individual probabilities up to and including the given value of x in the calculation for $P(X \leqslant x)$.

For the binomial distribution $X \sim B(n, p)$ there are tables in the formula book giving $P(X \leqslant x)$ for various values of n and p.

An extract from the tables is shown below:

$p =$	0.05	0.10	0.15	0.20	0.25	0.30
$n = 5, x = 0$	0.7738	0.5905	0.4437	0.3277	0.2373	0.1681
1	0.9774	0.9185	0.8352	0.7373	0.6328	0.5282
2	0.9988	0.9914	0.9734	0.9421	0.8965	0.8369
3	1.0000	0.9995	0.9978	0.9933	0.9844	0.9692
4	1.0000	1.0000	0.9999	0.9997	0.9990	0.9976
$n = 6, x = 0$	0.7351	0.5314	0.3771	0.2621	0.1780	0.1176

For the binomial distribution $X \sim B(5, 0.3)$, this tells you that $P(X \leqslant 2) = 0.8369$ to 4 d.p.

You can also use the binomial cumulative probability function on your calculator to find $P(X \leqslant x)$ for any values of x, n and p.

Example 6

The random variable $X \sim B(20, 0.4)$. Find:

a $P(X \leq 7)$ **b** $P(X < 6)$ **c** $P(X \geq 15)$

a $P(X \leq 7) = 0.4159$	Use $n = 20$, $p = 0.5$ and $x = 7$. You can use tables or your calculator.
b $P(X < 6) = P(X \leq 5)$ $= 0.1256$	**Online** Use the binomial **cumulative distribution** function on your calculator. You want to find $P(X \leq 7)$, not $P(X = 7)$. On some calculators, this is labelled 'Binomial CD'.
c $P(X \geq 15) = 1 - P(X \leq 14)$ $= 1 - 0.9984$ $= 0.0016$	X can only take whole number values, so $P(X < 6) = P(X \leq 5)$.

When questions are set in context there are different forms of words that can be used to ask for probabilities. The correct interpretation of these phrases is critical, especially when dealing with cumulative probabilities. The table below gives some examples.

Phrase	Means	Calculation
... greater than 5 ...	$X > 5$	$1 - P(X \leq 5)$
... no more than 3 ...	$X \leq 3$	$P(X \leq 3)$
... at least 7 ...	$X \geq 7$	$1 - P(X \leq 6)$
... fewer than 10 ...	$X < 10$	$P(X \leq 9)$
... at most 8 ...	$X \leq 8$	$P(X \leq 8)$

Example 7

A spinner is designed so that the probability it lands on red is 0.3. Jane has 12 spins. Find the probability that Jane obtains:

a no more than 2 reds

b at least 5 reds.

Jane decides to use this spinner for a class competition. She wants the probability of winning a prize to be < 0.05. Each member of the class will have 12 spins and the number of reds will be recorded.

c Find how many reds are needed to win a prize.

Let X = the number of reds in 12 spins.
$X \sim B(12, 0.3)$

a $P(X \leqslant 2) = 0.2528$

'no more than 2' means $X \leqslant 2$.

b $P(X \geqslant 5) = 1 - P(X \leqslant 4)$
$\quad\quad\quad = 1 - 0.7237$
$\quad\quad\quad = 0.2763$

'at least 5' means $X \geqslant 5$.

Form a probability statement to represent the condition for winning a prize.

c Let r = the smallest number of reds needed to win a prize.
Require: $P(X \geqslant r) < 0.05$

From tables:
$P(X \leqslant 5) = 0.8822$
$P(X \leqslant 6) = 0.9614$
$P(X \leqslant 7) = 0.9905$

So: $P(X \leqslant 6) = 0.9614$ implies that
$\quad P(X \geqslant 7) = 1 - 0.9614$
$\quad\quad\quad\quad\quad = 0.0386 < 0.05$

So 7 or more reds will win a prize.

Problem-solving

When you are looking for the first probability that is greater or less than a given threshold, it is sometimes quicker to look on the tables rather than use your calculator.

Since $x = 6$ gives the first value > 0.95, use this probability and find $r = 7$.

Always make sure that your final answer is related back to the context of the original question.

Online Explore the cumulative probabilities for the binomial distribution for this example using technology.

Exercise 6C

1 The random variable $X \sim B(9, 0.2)$. Find:
   a $P(X \leqslant 4)$ **b** $P(X < 3)$ **c** $P(X \geqslant 2)$ **d** $P(X = 1)$

2 The random variable $X \sim B(20, 0.35)$. Find:
   a $P(X \leqslant 10)$ **b** $P(X > 6)$
   c $P(X = 5)$ **d** $P(2 \leqslant X \leqslant 7)$

Hint $P(2 \leqslant X \leqslant 7) = P(X \leqslant 7) - P(X \leqslant 1)$

3 The random variable $X \sim B(40, 0.47)$. Find:
   a $P(X < 20)$ **b** $P(X > 16)$
   c $P(11 \leqslant X \leqslant 15)$ **d** $P(10 < X < 17)$

Watch out For questions **3** and **4** the values of n and p aren't given in the tables. Use the binomial cumulative distribution function on your calculator.

4 The random variable $X \sim B(37, 0.65)$. Find:
   a $P(X > 20)$ **b** $P(X \leqslant 26)$
   c $P(15 \leqslant X < 20)$ **d** $P(X = 23)$

(P) **5** Eight fair coins are tossed and the total number of heads showing is recorded.
   Find the probability of:
   a no heads **b** at least 2 heads **c** more heads than tails.

(P) **6** For a particular type of plant 25% have blue flowers. A garden centre sells these plants in trays of 15 plants of mixed colours. A tray is selected at random.
Find the probability that the number of plants with blue flowers in this tray is:

    a exactly 4

    b at most 3

    c between 3 and 6 (inclusive).

(E/P) **7** The random variable $X \sim B(50, 0.40)$. Find:

    a the largest value of k such that $P(X \leqslant k) < 0.05$ **(1 mark)**

    b the smallest number r such that $P(X > r) < 0.01$.   **(2 marks)**

(E/P) **8** The random variable $X \sim B(40, 0.10)$. Find:

    a the largest value of k such that $P(X < k) < 0.02$ **(1 mark)**

    b the smallest number r such that $P(X > r) < 0.01$ **(2 marks)**

    c $P(k \leqslant X \leqslant r)$.   **(2 marks)**

(E/P) **9** In a town, 30% of residents listen to the local radio.
Ten residents are chosen at random.
X = the number of these 10 residents that listen to the local radio.

    a Suggest a suitable distribution for X and comment on any necessary assumptions.   **(2 marks)**

    b Find the probability that at least half of these 10 residents listen to local radio.   **(2 marks)**

    c Find the smallest value of s so that $P(X \geqslant s) < 0.01$.   **(2 marks)**

(E/P) **10** A factory produces a component for the motor trade and 5% of the components are defective. A quality control officer regularly inspects a random sample of 50 components. Find the probability that the next sample contains:

    a fewer than 2 defectives   **(1 mark)**

    b more than 5 defectives.   **(2 marks)**

    The officer will stop production if the number of defectives in the sample is greater than a certain value d. Given that the officer stops production less than 5% of the time:

    c find the smallest value of d.   **(2 marks)**

Mixed exercise 6

(E) **1** The random variable X has probability function

$$P(X = x) = \frac{x}{21} \qquad x = 1, 2, 3, 4, 5, 6.$$

    a Construct a table giving the probability distribution of X.

    b Find $P(2 < X \leqslant 5)$.

(E) **2** The discrete random variable X has the probability distribution shown.

x	-2	-1	0	1	2	3
$P(X = x)$	0.1	0.2	0.3	r	0.1	0.1

    a Find r. **(1 mark)**

    b Calculate $P(-1 \leqslant x < 2)$. **(2 marks)**

(E/P) **3** The random variable X has probability function

$$P(X = x) = \frac{(3x - 1)}{26} \qquad x = 1, 2, 3, 4.$$

    a Construct a table giving the probability distribution of X. **(2 marks)**

    b Find $P(2 < X \leqslant 4)$. **(2 marks)**

(E) **4** Sixteen counters are numbered 1 to 16 and placed in a bag. One counter is chosen at random and the number, X, recorded.

    a Write down one condition on selecting a counter if X is to be modelled as a discrete uniform distribution. **(1 mark)**

    b Find:

       i $P(X = 5)$ **(1 mark)**

       ii $P(X \text{ is prime})$ **(2 marks)**

      iii $P(3 \leqslant X < 11)$ **(2 marks)**

(E/P) **5** The random variable Y has probability function

$$P(Y = y) = \frac{y}{k} \qquad y = 1, 2, 3, 4, 5$$

    a Find the value of k. **(2 marks)**

    b Construct a table giving the probability distribution of Y. **(2 marks)**

    c Find $P(Y > 3)$. **(1 mark)**

(E/P) **6** Stuart rolls a biased dice four times. $P(\text{six}) = \frac{1}{4}$. The random variable T represents the number of times he rolls a six.

    a Construct a table giving the probability distribution of T. **(3 marks)**

    b Find $P(T < 3)$. **(2 marks)**

    He rolls the dice again, this time recording the number of rolls required to roll a six. He rolls the dice a maximum of five times. Let the random variable S stand for the number of times he rolls the dice.

    c Construct a table giving the probability distribution of S. **(3 marks)**

    d Find $P(S > 2)$. **(2 marks)**

(E) **7** The discrete random variable $X \sim B(30, 0.73)$. Find:

    a $P(X = 20)$ **(1 mark)**

    b $P(X \leqslant 13)$ **(1 mark)**

    c $P(11 < X \leqslant 25)$ **(2 marks)**

(P) **8** A coin is biased so that the probability of a head is $\frac{2}{3}$. The coin is tossed repeatedly.
Find the probability that:

a the first tail will occur on the sixth toss

b in the first 8 tosses there will be exactly 2 tails.

(P) **9** Records kept in a hospital show that 3 out of every 10 patients who visit the accident and
emergency department have to wait more than half an hour. Find, to 3 decimal places, the
probability that of the first 12 patients who come to the accident and emergency department:

a none

b more than 2

will have to wait more than half an hour.

(E/P) **10 a** State clearly the conditions under which it is appropriate to assume that a random
variable has a binomial distribution. **(2 marks)**

A door-to-door canvasser tries to persuade people to have a certain type of double glazing
installed. The probability that his canvassing at a house is successful is 0.05.

b Find the probability that he will have at least 2 successes out of the first 10 houses
he canvasses. **(2 marks)**

c Calculate the smallest number of houses he must canvass so that the probability
of his getting at least one success exceeds 0.99. **(4 marks)**

(P) **11** A completely unprepared student is given a true/false-type test with 10 questions.
Assuming that the student answers all the questions at random:

a find the probability that the student gets all the answers correct.

It is decided that a pass will be awarded for 8 or more correct answers.

b Find the probability that the student passes the test.

(P) **12** A six-sided die is biased. When the die is thrown the number 5 is twice as likely to appear as
any other number. All the other faces are equally likely to appear. The die is thrown repeatedly.
Find the probability that:

a the first 5 will occur on the sixth throw

b in the first eight throws there will be exactly three 5s.

(E/P) **13** A manufacturer produces large quantities of plastic chairs. It is known from previous records
that 15% of these chairs are green. A random sample of 10 chairs is taken.

a Define a suitable distribution to model the number of green chairs in this sample. **(1 mark)**

b Find the probability of at least 5 green chairs in this sample. **(3 marks)**

c Find the probability of exactly 2 green chairs in this sample. **(3 marks)**

(E/P) **14** A bag contains a large number of beads of which 45% are yellow. A random sample of
20 beads is taken from the bag. Use the binomial distribution to find the probability that the
sample contains:

a fewer than 12 yellow beads **(2 marks)**

b exactly 12 yellow beads. **(3 marks)**

15 An archer hits the bullseye with probability 0.6. She shoots 20 arrows at a time.

    a Find the probability that she hits the bullseye with at least 50% of her arrows. **(3 marks)**

    She shoots 12 sets of 20 arrows.

    b Find the probability that she hits the bullseye with at least 50% of her arrows in 7 of the 12 sets of arrows. **(2 marks)**

    c Find the probability that she hits the bullseye with at least 50% of her arrows in fewer than 6 sets of arrows. **(2 marks)**

Challenge

A driving theory test has 50 questions. Each question has four answers, of which only one is correct.

Annabelle is certain she got 32 answers correct, but she guessed the remaining answers. She needs to get 43 correct answers to pass the test.

Find the probability that Annabelle passed the test.

Summary of key points

1 A **probability distribution** fully describes the probability of any outcome in the sample space.

2 The sum of the probabilities of all outcomes of an event add up to 1. For a random variable X, you can write $\sum P(X = x) = 1$ for all x.

3 You can model X with a **binomial distribution**, **B(n, p)**, if:
  - there are a fixed number of trials, n
  - there are two possible outcomes (success or failure)
  - there is a fixed probabillity of success, p
  - the trials are independent of each other.

4 If a random variable X has the binomial distribution B(n, p) then its probability mass function is given by
$$P(X = r) = \binom{n}{r}p^r(1 - p)^{n-r}$$

Hypothesis testing

Objectives

After completing this chapter you should be able to:

● Understand the language and concept of hypothesis
  testing   → **pages 99–101**

● Understand that a sample is used to make an inference
  about a population   → **pages 99–101**

● Find critical values of a binomial distribution using tables
                                                       → **pages 101–105**

● Carry out a one-tailed test for the proportion of the binomial
  distribution and interpret the results   → **pages 105–107**

● Carry out a two-tailed test for the proportion of the binomial
  distribution and interpret the results   → **pages 107–109**

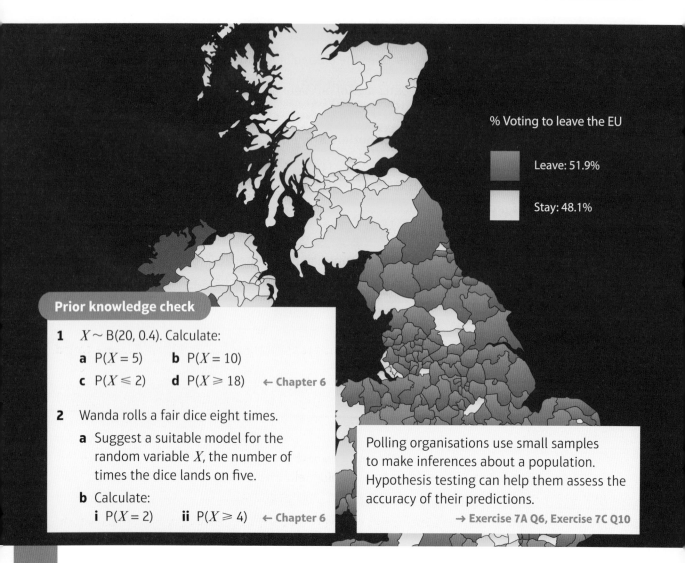

% Voting to leave the EU

Leave: 51.9%

Stay: 48.1%

Prior knowledge check

1 $X \sim B(20, 0.4)$. Calculate:
   a $P(X = 5)$ **b** $P(X = 10)$
   c $P(X \leqslant 2)$ **d** $P(X \geqslant 18)$ ← **Chapter 6**

2 Wanda rolls a fair dice eight times.
   a Suggest a suitable model for the
       random variable X, the number of
       times the dice lands on five.
   b Calculate:
       i $P(X = 2)$ **ii** $P(X \geqslant 4)$ ← **Chapter 6**

Polling organisations use small samples
to make inferences about a population.
Hypothesis testing can help them assess the
accuracy of their predictions.
                    → **Exercise 7A Q6, Exercise 7C Q10**

7.1 Hypothesis testing

A hypothesis is a statement made about the value of a **population parameter**. You can test a hypothesis about a population by carrying out an experiment or taking a sample from the population.

Links In this chapter the population parameter you will be testing will be the probability, p, in a binomial distribution $B(n, p)$. ← **Chapter 6**

The result of the experiment or the statistic that is calculated from the sample is called the **test statistic**.

In order to carry out the test, you need to form two hypotheses:

- **The null hypothesis, H_0, is the hypothesis that you assume to be correct.**
- **The alternative hypothesis, H_1, tells you about the parameter if your assumption is shown to be wrong.**

Notation In this chapter you should always give H_0 and H_1 in terms of the population parameter p.

Example 1

John wants to see whether a coin is unbiased or whether it is biased towards coming down heads. He tosses the coin 8 times and counts the number of times, X, that it lands head uppermost.

a Describe the test statistic.

b Write down a suitable null hypothesis.

c Write down a suitable alternative hypothesis.

a The test statistic is X (the number of heads in 8 tosses).	The test statistic is calculated from the sample or experiment.
b If the coin is unbiased the probability of a coin landing heads is 0.5 so H_0: $p = 0.5$ is the null hypothesis.	You always write the null hypothesis in the form H_0: $p = \ldots$
c If the coin is biased towards coming down heads then the probability of landing heads will be greater than 0.5. H_1: $p > 0.5$ is the alternative hypothesis.	If you were testing the coin for bias towards tails your alternative hypothesis would be H_1: $p < 0.5$. If you were testing the coin for bias in either direction your alternative hypothesis would be H_1: $p \neq 0.5$.

- **Hypothesis tests with alternative hypotheses in the form H_1: $p < \ldots$ and H_1: $p > \ldots$ are called one-tailed tests.**

- **Hypothesis tests with an alternative hypothesis in the form H_1: $p \neq \ldots$ are called two-tailed tests.**

Hint You can think of a two-tailed test such as H_1: $p \neq 0.5$ as **two tests**, H_1: $p > 0.5$ or $p < 0.5$.

To carry out a hypothesis test you **assume the null hypothesis is true**, then consider how likely the observed value of the test statistic was to occur. If this likelihood is less than a given threshold, called the **significance level** of the test, then you reject the null hypothesis.

Typically the significance level for a hypothesis test will be 10%, 5% or 1% but you will be told which level to use in the question.

Example (2)

An election candidate believes she has the support of 40% of the residents in a particular town. A researcher wants to test, at the 5% significance level, whether the candidate is over-estimating her support. The researcher asks 20 people whether they support the candidate or not. 3 people say that they do.

a Write down a suitable test statistic.

b Write down two suitable hypotheses.

c Explain the condition under which the null hypothesis would be rejected.

---

a The test statistic is the number of people who say they support the candidate.

b H_0: $p = 0.4$ H_1: $p < 0.4$

c The null hypothesis will be rejected if the probability of 3 or fewer people saying they support the candidate is less than 5%, given that $p = 0.4$.

This is a one-tailed test – the researcher wants to see if the candidate is **over-estimating** her support – if she is then the actual proportion of residents who support her will be **less than** 40%.

Watch out   You are testing to see whether the actual probability is **less** than 0.4, so you would need to calculate the probability that the observed value of the test statistic is 3 or **fewer**.

Exercise (7A)

1 a Explain what you understand by a hypothesis test.

   b Define a null hypothesis and an alternative hypothesis and state the symbols used for each.

   c Define a test statistic.

2 For each of these hypotheses, state whether the hypotheses given describe a one-tailed or a two-tailed test:

   a H_0: $p = 0.8$, H_1: $p > 0.8$ **b** H_0: $p = 0.6$, H_1: $p \neq 0.6$ **c** H_0: $p = 0.2$, H_1: $p < 0.2$

3 Dmitri wants to see whether a dice is biased towards the value 6. He throws the dice 60 times and counts the number of sixes he gets.

   Hint   If the dice is biased towards 6 then the probability of landing on 6 will be greater than $\frac{1}{6}$.

   a Describe the test statistic.

   b Write down a suitable null hypothesis to test this dice.

   c Write down a suitable alternative hypothesis to test this dice.

(P) **4** Shell wants to test to see whether a coin is biased. She tosses the coin 100 times and counts the number of times she gets a head. Shell says that her test statistic is the probability of the coin landing on heads.

   a Explain the mistake that Shell has made and state the correct test statistic for her test.

   b Write down a suitable null hypothesis to test this coin.

   c Write down a suitable alternative hypothesis to test this coin.

P **5** In a manufacturing process the proportion (p) of faulty articles has been found, from long experience, to be 0.1.

A sample of 100 articles from a new manufacturing process is tested, and 8 are found to be faulty.

The manufacturers wish to test at the 5% level of significance whether or not there has been a reduction in the proportion of faulty articles.

a Suggest a suitable test statistic.

b Write down two suitable hypotheses.

c Explain the condition under which the null hypothesis is rejected.

P **6** Polls show that 55% of voters support a particular political candidate. A newspaper releases information showing that the candidate avoided paying taxes the previous year. Following the release of the information, a polling company asked 20 people whether they support the candidate. 7 people said that they did. The polling company wants to test at the 2% level of significance whether the level of support for the candidate has reduced.

a Write down a suitable test statistic.

b Write down two suitable hypotheses.

c Explain the condition under which the null hypothesis would be accepted.

7.2 Finding critical values

When you carry out a hypothesis test, you need to be able to calculate the probability of the test statistic taking particular values given that the null hypothesis is true.

In this chapter you will assume that the test statistic can be modelled by a binomial distribution. You will use this to calculate probabilities and find critical regions.

- **A critical region is a region of the probability distribution which, if the test statistic falls within it, would cause you to reject the null hypothesis.**

A test statistic is modelled as B(10, p) and a hypothesis test at the 5% significance level uses H_0: $p = 0.4$, H_1: $p > 0.4$. Assuming H_0 to be true, X has the following distribution: $X \sim$ B(10, 0.4):

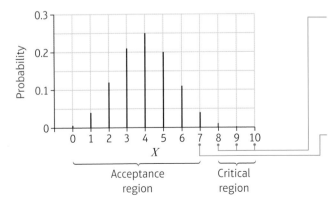

$P(X = 10) = 0.0001$, $P(X = 9) = 0.0016$ and $P(X = 8) = 0.0106$. Hence $P(X \geqslant 8) = 0.0123$. This is less than the significance level of 5%.
A test statistic of 10, 9 or 8 would lead to the null hypothesis being rejected.

$P(X = 7) = 0.0425$. Adding this probability to $P(X \geqslant 8)$ takes the probability over 0.05 so a test statistic of 7 or less would lead to the null hypothesis being accepted.

■ **The critical value is the first value to fall inside of the critical region.**

In this example, the critical value is 8 and the critical region is 8, 9 or 10. 7 falls in the **acceptance region**, the region where we accept the null hypothesis.

The critical value and hence the critical region can be determined from binomial distribution tables, or by finding cumulative binomial probabilities using your calculator.

In the $n = 10$ table, the critical region is found by looking for a probability such that $P(X \geqslant x) < 0.05$.

$p =$	0.20	0.25	0.30	0.35	0.40	0.45	0.50
$n = 10, x = 0$	0.1074	0.0563	0.0282	0.0135	0.0060	0.0025	0.0010
1	0.3758	0.2440	0.1493	0.0860	0.0464	0.0233	0.0107
2	0.6778	0.5256	0.3828	0.2616	0.1673	0.0996	0.0547
3	0.8791	0.7759	0.6496	0.5138	0.3823	0.2660	0.1719
4	0.9672	0.9219	0.8497	0.7515	0.6331	0.5044	0.3770
5	0.9936	0.9803	0.9527	0.9051	0.8338	0.7384	0.6230
6	0.9991	0.9965	0.9894	0.9740	0.9452	0.8980	0.8281
7	0.9999	0.9996	0.9984	0.9952	0.9877	0.9726	0.9453
8	1.0000	1.0000	0.9999	0.9995	0.9983	0.9955	0.9893
9	1.0000	1.0000	1.0000	1.0000	0.9999	0.9997	0.9990

$P(X \leqslant 6) = 0.9452$ so
$P(X \geqslant 7) = 0.0548$ (>0.05)

$P(X \leqslant 7) = 0.9877$ so
$P(X \geqslant 8) = 0.0132$ (<0.05)

This shows that $x = 7$ is not extreme enough to lead to the rejection of the null hypothesis but that $x = 8$ is. Hence $x = 8$ is the critical value and $x \geqslant 8$ is the critical region.

The probability of the test statistic falling within the critical region, given that H_0 is true is 0.0132 or 1.32%. This is sometimes called the **actual significance level** of the test.

■ **The actual significance level of a hypothesis test is the probability of incorrectly rejecting the null hypothesis.**

Watch out The threshold probability for your test (1%, 5%, 10%) is often referred to as the level of significance for your test. This might be different from the actual significance level, which is the probability that your test statistic would fall within the critical region even if H_0 is true.

Example **3**

A single observation is taken from a binomial distribution B(6, p). The observation is used to test H_0: $p = 0.35$ against H_1: $p > 0.35$.

a Using a 5% level of significance, find the critical region for this test.

b State the actual significance level of this test.

a Assume H_0 is true then $X \sim B(6, 0.35)$

$P(X \geqslant 4) = 1 - P(X \leqslant 3) = 1 - 0.8826$

$= 0.1174$

$P(X \geqslant 5) = 1 - P(X \leqslant 4) = 1 - 0.9777$

$= 0.0223$

The critical region is 5 or 6.

Use tables or your calculator to find the first value of x for which $P(X \geqslant x) < 0.05$.

$P(X \geqslant 4) > 0.05$ but $P(X \geqslant 5) < 0.05$ so 5 is the critical value.

Online Find the critical region using your calculator.

b The actual significance level is the probability of incorrectly rejecting the null hypothesis:

P(reject null hypothesis) = P($X \geq 5$)

= 0.0223

= 2.23%

This is the same as the probability that X falls within the critical region.

- **For a two-tailed test there are two critical regions: one at each end of the distribution.**

Example 4

A random variable X has binomial distribution B(40, p). A single observation is used to test H$_0$: $p = 0.25$ against H$_1$: $p \neq 0.25$.

a Using the 2% level of significance, find the critical region of this test. The probability in each tail should be as close as possible to 0.01.

b Write down the actual significance level of the test.

a Assume H$_0$ is true then $X \sim$ B(40, 0.25)

Consider the lower tail:

P($X \leq 4$) = 0.0160

P($X \leq 3$) = 0.0047

Consider the upper tail:

P($X \geq 19$) = 1 − P($X \leq 18$) = 1 − 0.9983

= 0.0017

P($X \geq 18$) = 1 − P($X \leq 17$) = 1 − 0.9884

= 0.0116

The critical regions are $0 \leq X \leq 3$ and $18 \leq X \leq 40$.

b The actual significance level is 0.0047 + 0.0116 = 0.0163 = 1.63%

P($X \leq 3$) is closest to 0.01 so 3 is the critical value for this tail.

Watch out Read the question carefully. Even though P($X \geq 18$) is greater than 0.01 it is still the closest value to 0.01. The critical value for this tail is 18.

Online Use technology to explore the locations of the critical values for each tail in this example.

Exercise 7B

1 Explain what you understand by the following terms:

a critical value   b critical region   c acceptance region.

2 A test statistic has a distribution B(10, p). Given that H$_0$: $p = 0.2$, H$_1$: $p > 0.2$, find the critical region for the test using a 5% significance level.

3 A random variable has a distribution B(20, p). A single observation is used to test H$_0$: $p = 0.15$ against H$_1$: $p < 0.15$. Using a 5% level of significance, find the critical region of this test.

(E) **4** A random variable has distribution $B(20, p)$. A single observation is used to test $H_0: p = 0.4$ against $H_1: p \neq 0.4$.

   a Using the 5% level of significance, find the critical region of this test. **(3 marks)**

   b Write down the actual significance level of the test. **(1 mark)**

5 A test statistic has a distribution $B(20, p)$.
Given that $H_0: p = 0.18$, $H_1: p < 0.18$, find the critical region for the test using a 1% level of significance.

> **Watch out** These probabilities are not found in statistical tables. You can use your calculator to find cumulative probabilities for $B(n, p)$ with any values of n and p.

(E/P) **6** A random variable has distribution $B(10, p)$. A single observation is used to test $H_0: p = 0.22$ against $H_1: p \neq 0.22$.

   a Using a 1% level of significance, find the critical region of this test. The probability in each tail should be as close as possible to 0.005. **(3 marks)**

   b Write down the actual significance level of the test. **(2 marks)**

(E/P) **7** A mechanical component fails, on average, 3 times out of every 10. An engineer designs a new system of manufacture that he believes reduces the likelihood of failure. He tests a sample of 20 components made using his new system.

   a Describe the test statistic. **(1 mark)**

   b State suitable null and alternative hypotheses. **(2 marks)**

   c Using a 5% level of significance, find the critical region for a test to check his belief, ensuring the probability is as close as possible to 0.05. **(3 marks)**

   d Write down the actual significance level of the test. **(1 mark)**

(E/P) **8** Seedlings come in trays of 36. On average, 12 seedlings survive to be planted on. A gardener decides to use a new fertiliser on the seedlings which she believes will improve the number that survive.

   a Describe the test statistic and state suitable null and alternative hypotheses. **(3 marks)**

   b Using a 10% level of significance, find the critical region for a test to check her belief. **(3 marks)**

   c State the probability of incorrectly rejecting H_0 using this critical region. **(1 mark)**

(E/P) **9** A restaurant owner notices that her customers typically choose lasagne one fifth of the time. She changes the recipe and believes this will change the proportion of customers choosing lasagne.

   a Suggest a model and state suitable null and alternative hypotheses. **(3 marks)**

   She takes a random sample of 25 customers.

   b Find, at the 5% level of significance, the critical region for a test to check her belief. **(4 marks)**

   c State the probability of incorrectly rejecting H_0. **(1 mark)**

A test statistic has binomial distribution B(50, p). Given that:

$H_0: p = 0.7$, $H_1: p \neq 0.7$:

a find the critical region for the test statistic such that the probability in each tail is close as possible to 5%.

Chloe takes two observations of the test statistic and finds that they both fall inside the critical region. Chloe decides to reject H_0.

b Find the probability that Chloe has incorrectly rejected H_0.

7.3 One-tailed tests

If you have to carry out a one-tailed hypothesis test you need to:

- Formulate a model for the test statistic
- Identify suitable null and alternative hypotheses
- Calculate the probability of the test statistic taking the observed value (or higher/lower), assuming the null hypothesis is true
- Compare this to the significance level
- Write a conclusion in the context of the question

Alternatively, you can find the critical region and see whether the observed value of the test statistic lies inside it.

Example 5

The standard treatment for a particular disease has a $\frac{2}{5}$ probability of success. A certain doctor has undertaken research in this area and has produced a new drug which has been successful with 11 out of 20 patients. The doctor claims that the new drug represents an improvement on the standard treatment.

Test, at the 5% significance level, the claim made by the doctor.

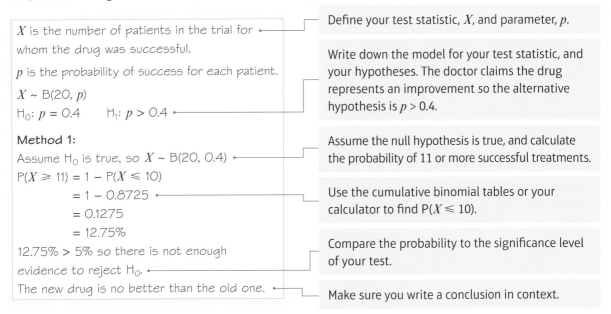

X is the number of patients in the trial for whom the drug was successful.

p is the probability of success for each patient.

$X \sim B(20, p)$

$H_0: p = 0.4$ $H_1: p > 0.4$

Method 1:

Assume H_0 is true, so $X \sim B(20, 0.4)$

$P(X \geq 11) = 1 - P(X \leq 10)$

           $= 1 - 0.8725$

           $= 0.1275$

           $= 12.75\%$

12.75% > 5% so there is not enough evidence to reject H_0.

The new drug is no better than the old one.

Define your test statistic, X, and parameter, p.

Write down the model for your test statistic, and your hypotheses. The doctor claims the drug represents an improvement so the alternative hypothesis is $p > 0.4$.

Assume the null hypothesis is true, and calculate the probability of 11 or more successful treatments.

Use the cumulative binomial tables or your calculator to find $P(X \leq 10)$.

Compare the probability to the significance level of your test.

Make sure you write a conclusion in context.

Method 2:

$P(X \geqslant 13) = 1 - P(X \leqslant 12) = 0.021$

$P(X \geqslant 12) = 1 - P(X \leqslant 11) = 0.0565$

The critical region is 13 or more.

Since 11 does not lie in the critical region, we accept H_0.

There is no evidence that the new drug is better than the old one.

Work out the critical region and see if 11 lies within it.

Unless you are specifically instructed as to which method to use, you can use the one you prefer.

Exercise 7C

1 A single observation, x, is taken from a binomial distribution $B(10, p)$ and a value of 5 is obtained. Use this observation to test H_0: $p = 0.25$ against H_1: $p > 0.25$ using a 5% significance level.

2 A random variable has distribution $X \sim B(10, p)$. A single observation of $x = 1$ is taken from this distribution. Test, at the 5% significance level, H_0: $p = 0.4$ against H_1: $p < 0.4$.

3 A single observation, x, is taken from a binomial distribution $B(20, p)$ and a value of 10 is obtained. Use this observation to test H_0: $p = 0.3$ against H_1: $p > 0.3$ using a 5% significance level.

4 A random variable has distribution $X \sim B(20, p)$. A single observation of $x = 3$ is taken from this distribution. Test, at the 1% significance level, H_0: $p = 0.45$ against H_1: $p < 0.45$.

5 A single observation, x, is taken from a binomial distribution $B(20, p)$ and a value of 2 is obtained. Use this observation to test H_0: $p = 0.28$ against H_1: $p < 0.28$ using a 5% significance level.

6 A random variable has distribution $X \sim B(8, p)$. A single observation of $x = 7$ is taken from this distribution. Test, at the 5% significance level, H_0: $p = 0.32$ against H_1: $p > 0.32$.

(P) 7 A dice used in playing a board game is suspected of not giving the number 6 often enough. During a particular game it was rolled 12 times and only one 6 appeared. Does this represent significant evidence, at the 5% level of significance, that the probability of a 6 on this dice is less than $\frac{1}{6}$?

(P) 8 The success rate of the standard treatment for patients suffering from a particular skin disease is claimed to be 68%.

a In a sample of n patients, X is the number for which the treatment is successful.
Write down a suitable distribution to model X. Give reasons for your choice of model.

A random sample of 10 patients receives the standard treatment and in only 3 cases was the treatment successful. It is thought that the standard treatment was not as effective as it is claimed.

b Test the claim at the 5% level of significance.

(E/P) 9 A plant germination method is successful on average 4 times out of every 10. A horticulturist develops a new technique which she believes will improve the number of plants that successfully germinate. She takes a random sample of 20 seeds and attempts to germinate them.

a Using a 5% level of significance, find the critical region for a test to check her belief. **(4 marks)**

b Of her sample of 20 plants, the horticulturalist finds that 14 have germinated. Comment on this observation in light of the critical region. **(2 marks)**

 10 A polling organisation claims that the support for a particular candidate is 35%. It is revealed that the candidate will pledge to support local charities if elected. The polling organisation think that the level of support will go up as a result. It takes a new poll of 50 voters.

a Describe the test statistic and state suitable null and alternative hypotheses. **(2 marks)**

b Using a 5% level of significance, find the critical region for a test to check the belief. **(4 marks)**

c In the new poll, 28 people are found to support the candidate. Comment on this observation in light of the critical region. **(2 marks)**

7.4 Two-tailed tests

A one-tailed test is used to test when it is claimed that the probability has either gone up, or gone down. A two-tailed test is used when it is thought that the probability has changed in either direction.

- **For a two-tailed test, halve the significance level at the end you are testing.**

You need to know which tail of the distribution you are testing. If the test statistic is $X \sim B(n, p)$ then the **expected** outcome is np. If the observed value, x, is lower than this then consider $P(X \leq x)$. If the observed value is higher than the expected value, then consider $P(X \geq x)$. In your exam it will usually be obvious which tail you should test.

Example 6

Over a long period of time it has been found that in Enrico's restaurant the ratio of non-vegetarian to vegetarian meals is 2 to 1. In Manuel's restaurant in a random sample of 10 people ordering meals, 1 ordered a vegetarian meal. Using a 5% level of significance, test whether or not the proportion of people eating vegetarian meals in Manuel's restaurant is different to that in Enrico's restaurant.

The proportion of people eating vegetarian meals at Enrico's is $\frac{1}{3}$.

X is the number of people in the sample at Manuel's who order vegetarian meals.

p is the probability that a randomly chosen person at Manuel's orders a vegetarian meal.

$H_0 : p = \frac{1}{3}$ $H_1 : p \neq \frac{1}{3}$

Significance level 5%

If H_0 is true $X \sim B(10, \frac{1}{3})$

Hypotheses. The test will be two-tailed as we are testing if they are different.

Method 1:

$P(X \leqslant 1) = P(X = 0) + P(X = 1)$

$\qquad = \left(\frac{2}{3}\right)^{10} + 10\left(\frac{2}{3}\right)^{9}\left(\frac{1}{3}\right)$

$\qquad = 0.017\,34... + 0.086\,70...$

$\qquad = 0.104 \text{ (3 s.f.)}$

$0.104 > 0.025$

There is insufficient evidence to reject H_0.

There is no evidence that proportion of vegetarian meals at Manuel's restaurant is different to Enrico's.

Method 2:

Let c_1 and c_2 be the two critical values.

$P(X \leqslant c_1) \leqslant 0.025$ and $P(x \geqslant c_2) \leqslant 0.025$

For the lower tail:

$P(X \leqslant 0) = 0.017\,341... < 0.025$

$P(X \leqslant 1) = 0.104\,04... > 0.025$

So $c_1 = 0$

For the upper tail:

$P(X \geqslant 6) = 1 - P(X \leqslant 5)$

$\qquad = 0.076\,56... > 0.025$

$P(X \geqslant 7) = 1 - P(X \leqslant 6)$

$\qquad = 0.019\,66... < 0.025$

So $c_2 = 7$

The observed value of 1 does not lie in the critical region so H_0 is not rejected. There is no evidence that the proportion of people eating vegetarian meals has changed.

The expected value would be $10 \times \frac{1}{3} = 3.333...$
The observed value, 1, is less than this so consider $P(X \leqslant 1)$.

You can calculate the probabilities long-hand, like this, or use your calculator with $p = 0.3333$.

We use 0.025 because the test is two-tailed.

Conclusion and what it means in context.

The probability in each critical region should be less than $0.05 \div 2 = 0.025$.

Use the cumulative binomial function on your calculator, with $n = 10$ and $p = 0.3333$.

Write down a value on either side of the boundary to show that you have determined the correct critical values.

Remember to write a conclusion in the context of the question.

Exercise 7D

1 A single observation, x, is taken from a binomial distribution B(30, p) and a value of 10 is obtained. Use this observation to test H_0: $p = 0.5$ against H_1: $p \neq 0.5$ using a 5% significance level.

2 A random variable has distribution $X \sim$ B(25, p). A single observation of $x = 10$ is taken from this distribution. Test, at the 10% significance level, H_0: $p = 0.3$ against H_1: $p \neq 0.3$.

3 A single observation, x, is taken from a binomial distribution B(10, p) and a value of 9 is obtained. Use this observation to test H_0: $p = 0.75$ against H_1: $p \neq 0.75$ using a 5% significance level.

4 A random variable has distribution $X \sim$ B(20, p). A single observation of $x = 1$ is taken from this distribution. Test, at the 1% significance level, H_0: $p = 0.6$ against H_1: $p \neq 0.6$.

(P) 5 A random variable has distribution $X \sim B(50, p)$. A single observation of $x = 4$ is taken from this distribution. Test, at the 2% significance level, H_0: $p = 0.02$ against H_1: $p \neq 0.02$.

Watch out Although the observed value of 4 appears to be small, the expected value of X is actually $50 \times 0.02 = 1$. You need to consider the upper tail of the distribution: $P(X \geq 4)$.

(P) 6 A coin is tossed 20 times, and lands on heads 6 times. Use a two-tailed test with a 5% significance level to determine whether there is sufficient evidence to conclude that the coin is biased.

(E/P) 7 The national proportion of people experiencing complications after having a particular operation in hospitals is 20%. A hospital decides to take a sample of size 20 from their records.

    a Find critical regions, at the 5% level of significance, to test whether or not their proportion of complications differs from the national proportion. The probability in each tail should be as close to 2.5% as possible. **(5 marks)**

    b State the actual significance level of the test. **(1 mark)**

    The hospital finds that 8 of their 20 patients experienced complications.

    c Comment on this finding in light of your critical regions. **(2 marks)**

(E/P) 8 A machine makes glass bowls and it is observed that one in ten of the bowls have hairline cracks in them. The production process is modified and a sample of 20 bowls is taken. 1 of the bowls is cracked. Test, at the 10% level of significance, the hypothesis that the proportion of cracked bowls has changed as a result of the change in the production process. State your hypotheses clearly. **(7 marks)**

(E/P) 9 Over a period of time, Agnetha has discovered that the carrots that she grows have a 25% chance of being longer than 7 cm. She tries a new type of fertiliser. In a random sample of 30 carrots, 13 are longer than 7 cm. Agnetha claims that the new fertiliser has changed the probability of a carrot being longer than 7 cm. Test Agnetha's claim at the 5% significance level. State your hypotheses clearly. **(7 marks)**

(E/P) 10 A standard blood test is able to diagnose a particular disease with probability 0.96. A manufacturer suggests that a cheaper test will have the same probability of success. It conducts a clinical trial on 75 patients. The new test correctly diagnoses 63 of these patients. Test the manufacturer's claim at the 10% level, stating your hypotheses clearly. **(7 marks)**

Mixed exercise 7

(E/P) 1 Mai commutes to work five days a week on a train. She does two journeys a day. Over a long period of time she finds that the train is late 20% of the time. A new company takes over the train service Mai uses. Mai thinks that the service will be late more often. In the first week of the new service the train is late 3 times. You may assume that the number of times the train is late in a week has a binomial distribution. Test, at the 5% level of significance, whether or not there is evidence that there is an increase in the number of times the train is late. State your hypothesis clearly. **(7 marks)**

E/P **2** A marketing company claims that Chestly cheddar cheese tastes better than Cumnauld cheddar cheese.

Five people chosen at random as they entered a supermarket were asked to say which they preferred. Four people preferred Chestly cheddar cheese.

Test, at the 5% level of significance, whether or not the manufacturer's claim is true. State your hypothesis clearly. **(7 marks)**

E/P **3** Historical information finds that nationally 30% of cars fail a brake test.

a Give a reason to support the use of a binomial distribution as a suitable model for the number of cars failing a brake test. **(1 mark)**

b Find the probability that, of 5 cars taking the test, all of them pass the brake test. **(2 marks)**

A garage decides to conduct a survey of their cars. A randomly selected sample of 10 of their cars is tested. Two of them fail the test.

c Test, at the 5% level of significance, whether or not there is evidence to support the suggestion that cars in this garage fail less than the national average. **(7 marks)**

E/P **4** The proportion of defective articles in a certain manufacturing process has been found from long experience to be 0.1.

A random sample of 50 articles was taken in order to monitor the production. The number of defective articles was recorded.

a Using a 5% level of significance, find the critical regions for a two-tailed test of the hypothesis that 1 in 10 articles has a defect. The probability in each tail should be as near 2.5% as possible. **(4 marks)**

b State the actual significance level of the above test. **(2 marks)**

Another sample of 20 articles was taken at a later date. Four articles were found to be defective.

c Test, at the 10% significance level, whether or not there is evidence that the proportion of defective articles has increased. State your hypothesis clearly. **(5 marks)**

E/P **5** It is claimed that 50% of women use Oriels powder. In a random survey of 20 women, 12 said they did not use Oriels powder.

Test, at the 5% significance level, whether or not there is evidence that the proportion of women using Oriels powder is 0.5. State your hypothesis carefully. **(6 marks)**

E/P **6** The manager of a superstore thinks that the probability of a person buying a certain make of computer is only 0.2.

To test whether this hypothesis is true the manager decides to record the make of computer bought by a random sample of 50 people who bought a computer.

a Find the critical region that would enable the manager to test whether or not there is evidence that the probability is different from 0.2. The probability of each tail should be as close to 2.5% as possible. **(4 marks)**

b Write down the significance level of this test. **(2 marks)**

15 people buy that certain make.

c Comment on this observation in light of your critical region. **(2 marks)**

7 a Explain what is meant by:

  i a hypothesis test **ii** a critical value **iii** an acceptance region. **(3 marks)**

Johan believes the probability of him being late to school is 0.2. To test this claim he counts the number of times he is late in a random sample of 20 days.

  b Find the critical region for a two-tailed test, at the 10% level of significance, of whether the probability he is late for school differs from 0.2. **(5 marks)**

  c State the actual significance level of the test. **(1 mark)**

Johan discovers he is late for school in 7 out of the 20 days.

  d Comment on whether Johan should accept or reject his claim that the probability he is late for school is 0.2. **(2 marks)**

 8 From the large data set, the likelihood of a day with either zero or trace amounts of rain in Hurn in June 1987 was 0.5.

Poppy believes that the likelihood of a rain-free day in 2015 has increased.

In June 2015 in Hurn, 21 days were observed as having zero or trace amounts of rain.

Using a 5% significance level, test whether or not there is evidence to support Poppy's claim. **(6 marks)**

 9 A single observation x is to be taken from a binomial distribution $B(30, p)$. This observation is used to test H_0: $p = 0.35$ against H_1: $p \neq 0.35$.

  a Using a 5% level of significance, find the critical region for this test. The probability of rejecting either tail should be as close as possible to 2.5%. **(3 marks)**

  b State the actual significance level of this test. **(2 marks)**

The actual value of X obtained is 4.

  c State a conclusion that can be drawn based on this value giving a reason for your answer. **(2 marks)**

 10 A pharmaceutical company claims that 85% of patients suffering from a chronic rash recover when treated with a new ointment.

A random sample of 20 patients with this rash is taken from hospital records.

  a Write down a suitable distribution to model the number of patients in this sample who recover when treated with the new ointment. **(2 marks)**

  b Given that the claim is correct, find the probability that the ointment will be successful for exactly 16 patients. **(2 marks)**

The hospital believes that the claim is incorrect and the percentage who will recover is lower. From the records an administrator took a random sample of 30 patients who had been prescribed the ointment. She found that 20 had recovered.

  c Stating your hypotheses clearly, test, at the 5% level of significance, the hospital's belief. **(6 marks)**

Large data set

You will need access to the large data set and spreadsheet software to answer these questions.

1 The proportion of days with a recorded daily mean temperature greater than 15 °C in Leuchars between May 1987 and October 1987 was found to be 0.163 (3 s.f.).

A meteorologist wants to use a randomly chosen sample of 10 days to determine whether the probability of observing a daily mean temperature greater than 15 °C has increased significantly between 1987 and 2015.

   a Using a significance level of 5%, determine the critical region for this test.

   b Select a sample of 10 days from the 2015 data for Leuchars, and count the number of days with a mean temperature of greater than 15 °C.

   c Use your observation and your critical region to make a conclusion.

2 From the large data set, in Beijing in 1987, 23% of the days from May to October had a daily mean air temperature greater than 25 °C. Using a sample of size 10 from the data for daily mean air temperature in Beijing in 2015, test, at the 5% significance level, whether the proportion of days with a mean air temperature greater than 25 °C increased between 1987 and 2015.

Summary of key points

1 The null hypothesis, H_0, is the hypothesis that you assume to be correct.

2 The alternative hypothesis, H_1, tells us about the parameter if your assumption is shown to be wrong.

3 Hypothesis tests with alternative hypotheses in the form $H_1: p < \ldots$ and $H_1: p > \ldots$ are called one-tailed tests.

4 Hypothesis tests with an alternative hypothesis in the form $H_1: p \neq \ldots$ are called two-tailed tests.

5 A critical region is a region of the probability distribution which, if the test statistic falls within it, would cause you to reject the null hypothesis.

6 The critical value is the first value to fall inside of the critical region.

7 The actual significance level of a hypothesis test is the probability of incorrectly rejecting the null hypothesis.

8 For a two-tailed test the critical region is split at either end of the distribution.

9 For a two-tailed test, halve the significance level at each end you are testing.

Review exercise

← Sections 1.3, 1.5, 2.1

(E) **1** A researcher is hired by a cleaning company to survey the opinions of employees on a proposed pension scheme. The company employs 55 managers and 495 cleaners.

  a Explain what is meant by a census and give one disadvantage of using it in this context. **(2)**

To collect data the researcher decides to give a questionnaire to the first 50 cleaners to leave at the end of the day.

  b State the sampling method used by the researcher. **(1)**

  c Give 2 reasons why this method is likely to produce biased results. **(2)**

  d Explain briefly how the researcher could select a sample of 50 employees using:
    i a systematic sample
    ii a stratified sample. **(2)**

← Sections 1.1, 1.2, 1.3

(E/P) **2** Data on the daily maximum relative humidity in Camborne during 1987 is gathered from the large data set. The daily maximum relative humidity, h, on the first five days in May is given below:

Day	1st	2nd	3rd	4th	5th
h	100	91	77	83	86

  a Explain what is meant by opportunity sampling and describe one limitation of the method in this context. **(2)**

  b Calculate an estimate for the mean daily maximum relative humidity for these five days. **(1)**

Joanna concludes that it is not likely to be misty in Camborne in May.

  c State, with reasons, whether your answer in part **b** supports Joanna's conclusion. **(2)**

(E) **3** Summarised below are the distances, to the nearest mile, travelled to work by a random sample of 120 commuters.

Distance, x miles	Number of commuters
$0 < x \leqslant 10$	10
$10 < x \leqslant 20$	19
$20 < x \leqslant 30$	43
$30 < x \leqslant 40$	25
$40 < x \leqslant 50$	8
$50 < x \leqslant 60$	6
$60 < x \leqslant 70$	5
$70 < x \leqslant 80$	3
$80 < x \leqslant 90$	1

  a For this distribution, use linear interpolation to estimate its median. **(3)**

The midpoint of each class was represented by x and its corresponding frequency f giving
$\Sigma fx = 3610$ and $\Sigma fx^2 = 141\,600$

  b Estimate the mean and standard deviation of this distribution. **(2)**

← Sections 2.1, 2.4

(E/P) **4** From the large data set, the daily total sunshine, s, in Leeming during June 1987 is recorded.

The data is coded using $x = 10s + 1$ and the following summary statistics are obtained.

$n = 30$ $\Sigma x = 947$ $S_{xx} = 33\,065.37$

Find the mean and standard deviation of the daily total sunshine. **(4)**

← Section 2.5

(E) **5** Children from schools A and B took part in a fun run for charity. The times, to the nearest minute, taken by the children from school A are summarised below.

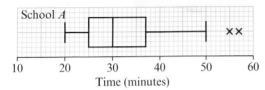

School A

Time (minutes)

a i Write down the time by which 75% of the children in school A had completed the run.

  ii State the name given to this value. **(2)**

b Explain what you understand by the two crosses (✗) on Figure 1. **(2)**

For school B the least time taken by any of the children was 25 minutes and the longest time was 55 minutes. The three quartiles were 30, 37 and 50 respectively.

c On graph paper, draw a box plot to represent the data from school B. **(3)**

d Compare and contrast these two box plots. **(2)**

←Sections 3.1, 3.2, 3.5

(E/P) **6** The histogram below shows the times taken, t, in minutes, by a group of people to swim 500 m.

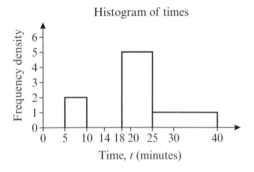

Histogram of times

Time, t (minutes)

a Copy and complete the frequency table and histogram for t. **(4)**

t	f
$5 \leqslant t < 10$	10
$10 \leqslant t < 14$	16
$14 \leqslant t < 18$	24
$18 \leqslant t < 25$	
$25 \leqslant t < 40$	

b Find the probability that a person chosen at random took longer than 20 minutes to swim 500 m. **(3)**

c Find an estimate of the mean time taken. **(2)**

d Find an estimate for the standard deviation of t. **(3)**

e Find an estimate for the median of t. **(2)**

←Sections 2.1, 2.4, 3.4

(E/P) **7** An ornithologist is collecting data on the lengths, in cm, of snowy owls. She displays the information in a histogram as shown below.

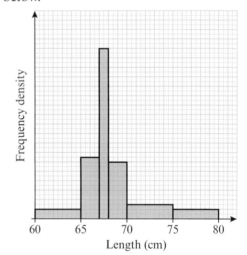

Length (cm)

Given that there are 26 owls in the 65 to 67 cm class, estimate the probability that an owl, chosen at random is between 63 and 73 cm long. **(4)**

← Section 3.4

(E/P) **8** The daily maximum temperature is recorded in a UK city during May 2015.

t	f
$10 \leqslant t < 13$	1
$13 \leqslant t < 16$	7
$16 \leqslant t < 19$	13
$19 \leqslant t < 22$	10

a Use linear interpolation to find an estimate for the 20% to 80% interpercentile range. **(3)**

b Draw a cumulative frequency diagram to display this data. **(3)**

c Use your diagram to estimate the 20% to 80% interpercentile range and compare your answer to part **a**. Which estimate is likely to be more accurate? **(3)**

d Estimate the number of days in May 2015 where the daily maximum temperature in this city is greater than 15 °C. **(2)**

← Sections 2.3, 3.3

(E) **9** A manufacturer stores drums of chemicals. During storage, evaporation take place. A random sample of 10 drums was taken and the time in storage, x weeks, and the evaporation loss, y ml, are shown in the table below.

x	3	5	6	8	10	12	13	15	16	18
y	36	50	53	61	69	79	82	90	88	96

a On graph paper, draw a scatter diagram to represent these data. **(3)**

b Give a reason to support fitting a regression model of the form $y = a + bx$ to these data. **(1)**

The equation of the regression line of y on x is $y = 29.02 + 3.9x$.

c Give an interpretation of the value of the gradient in the equation of the regression line. **(1)**

The manufacturer uses this model to predict the amount of evaporation that would take place after 19 weeks and after 35 weeks.

d Comment, with a reason, on the reliability of each of these predictions. **(2)**

← Sections 4.2

(E/P) **10** The table shows average monthly temperature, t (°C) and the number of ice creams, c, in 100s, a riverside snack barge sells each month.

t	7	8	10	45	14	17	20	21	15	13	9	5
c	4	7	13	27	30	35	42	41	36	24	9	3

The following statistics were calculated for the data on temperature: mean = 15.3, standard deviation = 10.2 (both correct to 3 s.f.)

An outlier is an observation which lies ±2 standard deviations from the mean.

a Show that $t = 45$ is an outlier. **(1)**

b Give a reason whether or not this outlier should be omitted from the data. **(1)**

This value is omitted from the data, and the equation of the regression line of c on t for the remaining data is calculated as $c = 2.81t - 13.3$.

c Give an interpretation of the value of 2.81 in this regression equation. **(1)**

d State, with a reason, why using the regression line to estimate the number of ice creams sold when the average monthly temperature is 2 °C would not be appropriate. **(2)**

← Sections 3.1, 4.2

(E) **11** A bag contains nine blue balls and three red balls. A ball is selected at random from the bag and its colour is recorded. The ball is not replaced. A second ball is selected at random and its colour is recorded.

a Draw a tree diagram to represent the information **(3)**

Find the probability that:

b the second ball selected is red **(1)**

c the balls are different colours. **(3)**

← Section 5.4

(E/P) **12** For events A and B, P(A but not B) = 0.32, P(B but not A) = 0.11 and P(A or B) = 0.65.

a Draw a Venn diagram to illustrate the complete sample space for the events A and B. **(3)**

b Write down the value of P(A) and the value of P(B). **(2)**

c Determine whether or not A and B are independent. **(2)**

← Sections 5.2, 5.3

(E) **13** A company assembles drills using components from two sources. Goodbuy supplies 85% of the components and

Amart supplies the rest. It is known that 3% of the components supplied by Goodbuy are faulty and 6% of those supplied by Amart are faulty.

a Represent this information on a tree diagram. **(2)**

An assembled drill is selected at random.

b Find the probability that it is not faulty. **(3)**

← **Section 5.4**

(E/P) **14** The Venn diagram shows the number of children who like Comics (C), Books (B) or Television (T).

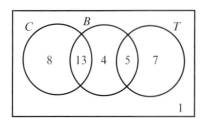

a Which two hobbies are mutually exclusive? **(1)**

b Determine whether the events 'likes comics' and 'likes books' are independent. **(3)**

← **Sections 5.2, 5.3**

(E/P) **15** A fair coin is tossed 4 times.

Find the probability that:

a an equal number of heads and tails occur **(3)**

b all the outcomes are the same **(2)**

c the first tail occurs on the third throw. **(2)**

← **Section 6.1**

(E/P) **16** The probability of a bolt being faulty is 0.3. Find the probability that in a random sample of 20 bolts there are:

a exactly 2 faulty bolts **(2)**

b more that 3 faulty bolts. **(2)**

These bolts are sold in bags of 20. John buys 10 bags.

c Find the probability that exactly 6 of these bags contain more than 3 faulty bolts. **(3)**

← **Sections 6.2, 6.3**

(E) **17** The random variable X has probability function

$$P(X = x) = \frac{(2x - 1)}{36} \quad x = 1, 2, 3, 4, 5, 6.$$

a Construct a table giving the probability distribution of X. **(3)**

Find

b $P(2 < X \leqslant 5)$ **(2)**

← **Section 6.1**

(E/P) **18** A fair five-sided spinner has sectors numbered from 1 to 5. The spinner is spun and the number showing, X, is recorded.

a State the distribution of X. **(1)**

The spinner is spun four times and the number of spins taken to get an odd number, Y, is recorded.

b Write down, in table form, the probability distribution of Y. **(3)**

c Find $P(Y > 2)$. **(2)**

← **Section 6.1**

(E) **19** The random variable $X \sim B(15, 0.32)$. Find:

a $P(X = 7)$ **b** $P(X \leqslant 4)$

c $P(X < 8)$ **d** $P(X \geqslant 6)$ **(4)**

← **Section 6.3**

(E/P) **20** A single observation is taken from a test statistic $X \sim B(40, p)$. Given that $H_0: p = 0.3$ and $H_1: p \neq 0.3$,

a Find the critical region for the test using a 2.5% significance level. (The probability in each tail should be as close as possible to 1.25 %) **(4)**

b State the probability of incorrectly rejecting the null hypothesis using this test. **(1)**

← **Section 7.2**

E/P **21** A drugs company claims that 75% of patients suffering from depression recover when treated with a new drug.

A random sample of 10 patients with depression is taken from a doctor's records.

a Write down a suitable distribution to model the number of patients in this sample who recover when treated with the new drug. **(1)**

Given that the claim is correct,

b find the probability that the treatment will be successful for exactly 6 patients. **(2)**

The doctor believes that the claim is incorrect and the percentage who will recover is lower. From her records she took a random sample of 20 patients who had been treated with the new drug. She found that 13 had recovered.

c Stating your hypotheses clearly, test, at the 5% level of significance, the doctor's belief. **(7)**

d From a sample of size 20, find the greatest number of patients who need to recover from the test in part **c** to be significant at the 1% level. **(3)**

← Sections 6.3, 7.1, 7.2, 7.3

E/P **22** Dhriti grows tomatoes. Over a period of time, she has found that there is a probability 0.3 of a ripe tomato having a diameter greater than 4 cm. Dhriti wants to test whether a new fertiliser increases the size of her tomatoes. She takes a sample of 40 ripe tomatoes that have been treated with the new fertiliser.

a Write down suitable hypotheses for her test. **(1)**

b Using a 5% significance level, find the critical region for her test. **(4)**

c Write down the actual significance level of the test. **(1)**

Dhriti finds that 18 out of the 40 tomatoes have a diameter greater than 4 cm.

d Comment on Dhriti's observed value in light of your answer to part **b**. **(2)**

← Sections 7.2, 7.3

Challenge

1 The Venn diagram shows the number of sports club members liking three different sports.

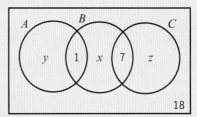

Given that there are 50 members in total, $P(C) = 3P(A)$ and $P(\text{not } B) = 0.76$, find the values of x, y and z. ← Section 5.2

2 A test statistic has binomial distribution $B(30, p)$. Given that $H_0: p = 0.65$, $H_1: p < 0.65$,

a find the critical region for the test statistic such that the probability is as close as possible to 10%.

William takes two observations of the test statistic and finds that they both fall inside the critical region. William decides to reject H_0.

b Find the probability that William has incorrectly rejected H_0. ← Sections 7.2, 7.3

8 Modelling in mechanics

Objectives

After completing this chapter you should be able to:

● Understand how the concept of a mathematical model
  applies to mechanics   → pages 119–120

● Understand and be able to apply some of the common
  assumptions used in mechanical models   → pages 120–122

● Know SI units for quantities and derived quantities used in
  mechanics   → pages 122–124

● Know the difference between scalar and vector quantities
                                                          → pages 125–127

Prior knowledge check

Give your answers correct to 3 s.f. where appropriate.

1   Solve these equations:

    a $5x^2 - 21x + 4 = 0$ **b** $6x^2 + 5x = 21$

    c $3x^2 - 5x - 4 = 0$ **d** $8x^2 - 18 = 0$

    ← GCSE Mathematics

2   Find the value of x and y in these right-angled
    triangles.

    a   **b**

    ← GCSE Mathematics

3   Convert:

    a $30\,\text{km h}^{-1}$ to cm s^{-1} **b** $5\,\text{g cm}^{-3}$ to kg m^{-3}

    ← GCSE Mathematics

4   Write in standard form:

    a 7 650 000   **b** 0.003 806

    ← GCSE Mathematics

Mathematical models can be used to
find solutions to **real-world problems** in
many everyday situations. If you model a
firework as a particle you can ignore the
effects of wind and air resistance.

8.1 Constructing a model

Mechanics deals with motion and the action of forces on objects. Mathematical models can be constructed to simulate real-life situations, but in many cases it is necessary to simplify the problem by making assumptions so that it can be described using equations or graphs in order to solve it.

The solution to a mathematical model needs to be interpreted in the context of the original problem. It is possible your model may need to be refined and your assumptions reconsidered.

This flow chart summarises the mathematical modelling process.

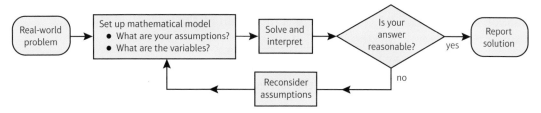

Example 1

The motion of a basketball as it leaves a player's hand and passes through the net can be modelled using the equation $h = 2 + 1.1x - 0.1x^2$, where h m is the height of the basketball above the ground and x m is the horizontal distance travelled.

a Find the height of the basketball: **i** when it is released   **ii** at a horizontal distance of 0.5 m.

b Use the model to predict the height of the basketball when it is at a horizontal distance of 15 m from the player.

c Comment on the validity of this prediction.

a **i** $x = 0$: $h = 2 + 0 + 0$ Height = 2 m	When the basketball is released at the start of the motion $x = 0$. Substitute $x = 0$ into the equation for h.
ii $x = 0.5$: $h = 2 + 1.1 \times 0.5 - 0.1 \times (0.5)^2$ Height = 2.525 m	Substitute $x = 0.5$ into the equation for h.
b $x = 15$: $h = 2 + 1.1 \times 15 - 0.1 \times (15)^2$ Height = −4 m	Substitute $x = 15$ into the equation for h.
c Height cannot be negative so the model is not valid when $x = 15$ m.	h represents the height of the basketball above the ground, so it is only valid if $h \geqslant 0$.

Exercise 8A

1 The motion of a golf ball after it is struck by a golfer can be modelled using the equation $h = 0.36x - 0.003x^2$, where h m is the height of the golf ball above the ground and x m is the horizontal distance travelled.

   a Find the height of the golf ball when it is:
     i struck   **ii** at a horizontal distance of 100 m.

   b Use the model to predict the height of the golf ball when it is 200 m from the golfer.

   c Comment on the validity of this prediction.

2 A stone is thrown into the sea from the top of a cliff. The height of the stone above sea level, h m at time t s after it is thrown can be modelled by the equation $h = -5t^2 + 15t + 90$.

    a Write down the height of the cliff above sea level.

    b Find the height of the stone:

       i when $t = 3$ **ii** when $t = 5$.

    c Use the model to predict the height of the stone after 20 seconds.

    d Comment on the validity of this prediction.

(P) **3** The motion of a basketball as it leaves a player's hand and passes through the net is modelled using the equation $h = 2 + 1.1x - 0.1x^2$, where h m is the height of the basketball above the ground and x m is the horizontal distance travelled.

    a Find the two values of x for which the basketball is exactly 4 m above the ground.

    This model is valid for $0 \leqslant x \leqslant k$, where k m is the horizontal distance of the net from the player. Given that the height of the net is 3 m:

    b Find the value of k.

    c Explain why the model is not valid for $x > k$.

(P) **4** A car accelerates from rest to 60 mph in 10 seconds. A quadratic equation of the form $d = kt^2$ can be used to model the distance travelled, d metres in time t seconds.

> **Problem-solving**
>
> Use the information given to work out the value of k.

    a Given that when $t = 1$ second the distance travelled by the car is 13.2 metres, use the model to find the distance travelled when the car reaches 60 mph.

    b Write down the range of values of t for which the model is valid.

(P) **5** The model for the motion of a golf ball given in question 1 is only valid when h is positive. Find the range of values of x for which the model is valid.

(P) **6** The model for the height of the stone above sea level given in question 2 is only valid from the time the stone is thrown until the time it enters the sea. Find the range of values of t for which the model is valid.

8.2 Modelling assumptions

Modelling assumptions can simplify a problem and allow you to analyse a real-life situation using known mathematical techniques. You need to understand the significance of different modelling assumptions and how they affect the calculations in a particular problem.

> **Watch out**
>
> Modelling assumptions can affect the validity of a model. For example, when modelling the landing of a commercial flight, it would not be appropriate to ignore the effects of wind and air resistance.

These are some common models and modelling assumptions that you need to know.

Model	Modelling assumptions
Particle – Dimensions of the object are negligible.	• mass of the object is concentrated at a single point • rotational forces and air resistance can be ignored
Rod – All dimensions but one are negligible, like a pole or a beam.	• mass is concentrated along a line • no thickness • rigid (does not bend or buckle)
Lamina – Object with area but negligible thickness, like a sheet of paper.	• mass is distributed across a flat surface
Uniform body – Mass is distributed evenly.	• mass of the object is concentrated at a single point at the geometrical centre of the body – the **centre of mass**
Light object – Mass of the object is small compared to other masses, like a string or a pulley.	• treat object as having zero mass • tension the same at both ends of a light string
Inextensible string – A string that does not stretch under load.	• acceleration is the same in objects connected by a taut inextensible string
Smooth surface	• assume that there is no friction between the surface and any object on it
Rough surface – If a surface is not smooth, it is rough.	• objects in contact with the surface experience a frictional force if they are moving or are acted on by a force
Wire – Rigid thin length of metal.	• treated as one-dimensional
Smooth and light pulley – all pulleys you consider will be smooth and light.	• pulley has no mass • tension is the same on either side of the pulley
Bead – Particle with a hole in it for threading on a wire or string.	• moves freely along a wire or string • tension is the same on either side of the bead
Peg – A support from which a body can be suspended or rested.	• dimensionless and fixed • can be rough or smooth as specified in question
Air resistance – Resistance experienced as an object moves through the air.	• usually modelled as being negligible
Gravity – Force of attraction between all objects. Acceleration due to gravity is denoted by g. $g = 9.8 \text{ m s}^{-2}$	• assume that all objects with mass are attracted towards the Earth • Earth's gravity is uniform and acts vertically downwards • g is constant and is taken as 9.8 m s^{-2}, unless otherwise stated in the question

Example 2

A mass is attached to a length of string which is fixed to the ceiling.
The mass is drawn to one side with the string taut and allowed to swing.
State the effect of the following assumptions on any calculations made using this model:

a The string is light and inextensible. **b** The mass is modelled as a particle.

a Ignore the mass of the string and any stretching effect caused by the mass.
b Ignore the rotational effect of any external forces that are acting on it, and the effects of air resistance.

Exercise 8B

1 A football is kicked by the goalkeeper from one end of the football pitch.
State the effect of the following assumptions on any calculations made using this model:
  a The football is modelled as a particle.   **b** Air resistance is negligible.

2 An ice puck is hit and slides across the ice.
State the effect of the following assumptions on any calculations made using this model:
  a The ice puck is modelled as a particle.   **b** The ice is smooth.

3 A parachute jumper wants to model her descent from an aeroplane to the ground. She models herself and her parachute as particles connected by a light inextensible string. Explain why this may not be a suitable modelling assumption for this situation.

4 A fishing rod manufacturer constructs a mathematical model to predict the behaviour of a particular fishing rod. The fishing rod is modelled as a light rod.
  a Describe the effects of this modelling assumption.
  b Comment on its validity in this situation.

5 Make a list of the assumptions you might make to create simple models of the following:
  a The motion of a golf ball after it is hit
  b The motion of a child on a sledge going down a snow-covered hill
  c The motion of two objects of different masses connected by a string that passes over a pulley
  d The motion of a suitcase on wheels being pulled along a path by its handle.

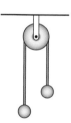

8.3 Quantities and units

The International System of Units, (abbreviated **SI** from the French, Système international d'unités) is the modern form of the metric system. These **base** SI units are most commonly used in mechanics.

Quantity	Unit	Symbol
Mass	kilogram	kg
Length/displacement	metre	m
Time	seconds	s

Watch out

A common misconception is that kilograms measure weight not mass. However, **weight** is a **force** which is measured in **newtons (N)**.

These **derived** units are compound units built from the base units.

Quantity	Unit	Symbol
Speed/velocity	metres per second	$m\,s^{-1}$
Acceleration	metres per second per second	$m\,s^{-2}$
Weight/force	newton	$N\ (= kg\,m\,s^{-2})$

You will encounter a variety of forces in mechanics. These **force diagrams** show some of the most common forces.

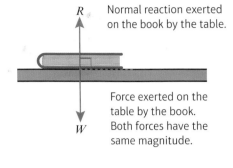

R — Normal reaction exerted on the book by the table.

W — Force exerted on the table by the book. Both forces have the same magnitude.

- The **weight** (or gravitational force) of an object acts vertically downwards.

- The **normal reaction** is the force which acts perpendicular to a surface when an object is in contact with the surface. In this example the normal reaction is due to the weight of the book resting on the surface of the table.

- The **friction** is a force which opposes the motion between two rough surfaces.

- If an object is being pulled along by a string, the force acting on the object is called the **tension** in the string.

- If an object is being pushed along using a light rod, the force acting on the object is called the **thrust** or **compression** in the rod.

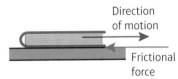

Direction of motion

Frictional force

Tension in string

Thrust or compression in rod

- **Buoyancy** is the upward force on a body that allows it to float or rise when submerged in a liquid. In this example buoyancy acts to keep the boat afloat in the water.

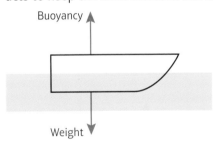

Buoyancy

Weight

- **Air resistance** opposes motion. In this example the weight of the parachutist acts vertically downwards and the air resistance acts vertically upwards.

Air resistance

Weight

Example 3

Write the following quantities in SI units.

a 4 km   **b** 0.32 grams   **c** $5.1 \times 10^6 \, km \, h^{-1}$

The SI unit of length is the metre and 1 km = 1000 m.

The SI unit of weight is the kg and 1 kg = 1000 g. The answer is given in standard form.

The SI unit of speed is $m \, s^{-1}$. Convert from $km \, h^{-1}$ to $m \, h^{-1}$ by multiplying by 100.

Convert from $m \, h^{-1}$ to $m \, s^{-1}$ by dividing by 60 × 60. The answer is given in standard form to 3 s.f.

a $4 \, km = 4 \times 1000 = 4000 \, m$

b $0.32 \, g = 0.32 \div 1000 = 3.2 \times 10^{-4} \, kg$

c $5.1 \times 10^6 \, km \, h^{-1} = 5.1 \times 10^6 \times 1000$
$= 5.1 \times 10^9 \, m \, h^{-1}$

$5.1 \times 10^9 \div (60 \times 60) = 1.42 \times 10^6 \, m \, s^{-1}$

Example 4

The force diagram shows an aircraft in flight.

Write down the names of the four forces shown on the diagram.

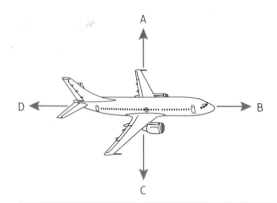

A upward thrust •	Also known as 'lift', this is the upward force that keeps the aircraft up in the air.
B forward thrust •	
C weight •	Also known as 'thrust', this is the force that propels the aircraft forward.
D air resistance •	
	This is the gravitational force acting downwards on the aircraft.

Exercise 8C

Also known as 'drag', this is the force that acts in the opposite direction to the forward thrust.

1 Convert to SI units:

  a $65\,\text{km}\,\text{h}^{-1}$ **b** $15\,\text{g}\,\text{cm}^{-2}$ **c** $30\,\text{cm}$ per minute

  d $24\,\text{g}\,\text{m}^{-3}$ **e** $4.5 \times 10^{-2}\,\text{g}\,\text{cm}^{-3}$ **f** $6.3 \times 10^{-3}\,\text{kg}\,\text{cm}^{-2}$

2 Write down the names of the forces shown in each of these diagrams.

  a A box being pushed along rough ground   **b** A man swimming through the water

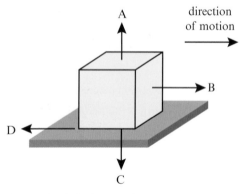

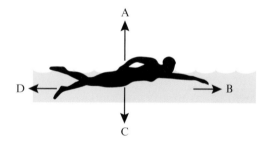

  c A toy duck being pulled along by a string   **d** A man sliding down a hill on a sledge

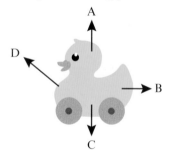

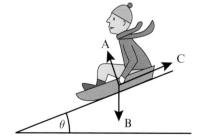

8.4 Working with vectors

- **A vector is a quantity which has both magnitude and direction.**

These are examples of **vector** quantities.

Quantity	Description	Unit
Displacement	distance in a particular direction	metre (m)
Velocity	rate of change of displacement	metres per second (m s^{-1})
Acceleration	rate of change of velocity	metres per second per second (m s^{-2})
Force/weight	described by magnitude, direction and point of application	newton (N)

- **A scalar quantity has magnitude only.**

These are examples of **scalar** quantities.

Quantity	Description	Unit
Distance	measure of length	metre (m)
Speed	measure of how quickly a body moves	metres per second (m s^{-1})
Time	measure of ongoing events taking place	second (s)
Mass	measure of the quantity of matter contained in an object	kilogram (kg)

Scalar quantities are always **positive**. When considering motion in a straight line (1-dimensional motion), **vector** quantities can be **positive** or **negative**.

Example 5

Fully describe the motion of the following particles:

	a	b	c	d
Velocity	+ve	+ve	−ve	−ve
Acceleration	+ve	−ve	−ve	+ve

Positive direction →

a The particle is moving to the right and its speed is increasing.

b The particle is moving to the right and its speed is decreasing.

c The particle is moving to the left and its speed is increasing.

d The particle is moving to the left and its speed is decreasing.

When the direction of the acceleration opposes the direction of motion, the particle is **slowing down**. This is also called **deceleration** or **retardation**.

You can describe vectors using **i**, **j** notation, where **i** and **j** are the unit vectors in the positive x and y directions.

Links When a vector is given in **i–j** notation you can:
- use Pythagoras' theorem to find its magnitude
- use trigonometry to work out its direction
  ← **Pure Year 1, Chapter 11**

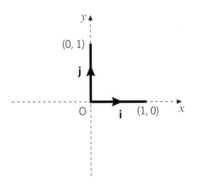

125

- **Distance is the magnitude of the displacement vector.**
- **Speed is the magnitude of the velocity vector.**

Example 6

The velocity of a particle is given by $\mathbf{v} = 3\mathbf{i} + 5\mathbf{j}$ m s⁻¹. Find:

a the speed of the particle

b the angle the direction of motion of the particle makes with the unit vector **i**.

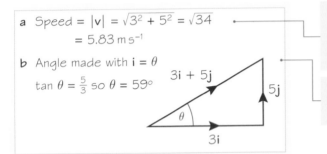

a Speed = $|\mathbf{v}| = \sqrt{3^2 + 5^2} = \sqrt{34}$
  = 5.83 m s⁻¹

b Angle made with $\mathbf{i} = \theta$
  $\tan \theta = \frac{5}{3}$ so $\theta = 59°$

The speed of the particle is the magnitude of the vector **v**. This is written as $|\mathbf{v}|$.
In general, if $\mathbf{v} = a\mathbf{i} + b\mathbf{j}$ then $|\mathbf{v}| = \sqrt{a^2 + b^2}$

Draw a diagram. The direction of **v** can be found using trigonometry.

Example 7

A man walks from A to B and then from B to C.

His displacement from A to B is $6\mathbf{i} + 4\mathbf{j}$ m.

His displacement from B to C is $5\mathbf{i} - 12\mathbf{j}$ m.

a What is the magnitude of the displacement from A to C?

b What is the total distance the man has walked in getting from A to C?

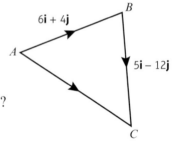

a $\overrightarrow{AC} = \overrightarrow{AB} + \overrightarrow{BC}$
  $\overrightarrow{AC} = \begin{pmatrix} 6 \\ 4 \end{pmatrix} + \begin{pmatrix} 5 \\ -12 \end{pmatrix} = \begin{pmatrix} 11 \\ -8 \end{pmatrix}$
  $|\overrightarrow{AC}| = \sqrt{11^2 + (-8)^2}$
  = 13.6 km

b Total distance = $|\overrightarrow{AB}| + |\overrightarrow{BC}|$
  $|\overrightarrow{AB}| = \sqrt{6^2 + 4^2} = 7.21$ km
  $|\overrightarrow{BC}| = \sqrt{5^2 + (-12)^2} = 13$ km
  total distance = $7.21 + 13 = 20.21$ km

This is the triangle law for vector addition. Write $\overrightarrow{AB}$ and $\overrightarrow{BC}$ in column vector form and add the **i** components and the **j** components to find $|\overrightarrow{AC}|$.
Use Pythagoras to work out the magnitude of $\overrightarrow{AC}$.

Note that the distance from A to C is not the same as the total distance travelled which is $|\overrightarrow{AB}| + |\overrightarrow{BC}|$.

Online Check your answers by entering the vectors directly into your calculator.

Exercise (8D)

1 A man walks from his home along a straight road to a shop with a speed of $2.1\,\mathrm{m\,s^{-1}}$ and walks home again at a speed of $1.8\,\mathrm{m\,s^{-1}}$.

He then jogs along a straight road from his home to the park with a speed of $2.7\,\mathrm{m\,s^{-1}}$ and returns home at a speed of $2.5\,\mathrm{m\,s^{-1}}$.

The park, the man's home and the shop all lie on a straight line, as shown in the diagram.

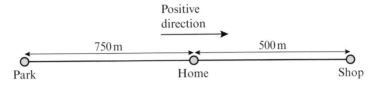

Taking the positive direction as shown in the diagram, state the man's:

a velocity on the journey from his home to the shop

b displacement from his home when he reaches the shop

c velocity on the journey from the shop to his home

d velocity on the journey from his home to the park

e displacement from his home when he reaches the park

f velocity on the journey from the park to his home.

2 The velocity of a car is given by $\mathbf{v} = 12\mathbf{i} - 10\mathbf{j}\,\mathrm{m\,s^{-1}}$. Find:

a the speed of the car

b the angle the direction of motion of the car makes with the unit vector $\mathbf{i}$.

3 The acceleration of a motorbike is given by $\mathbf{a} = 3\mathbf{i} - 4\mathbf{j}\,\mathrm{m\,s^{-2}}$. Find:

a the magnitude of the acceleration

b the angle the direction of the acceleration vector makes with the unit vector $\mathbf{j}$.

Problem-solving

Draw a sketch to help you find the direction. $\mathbf{j}$ acts in the positive y-direction, so the angle between $\mathbf{j}$ and the vector $3\mathbf{i} - 4\mathbf{j}$ will be obtuse.

4 A girl cycles from A to B and then from B to C.

The displacement from A to B is $10\mathbf{i} + 3\mathbf{j}\,\mathrm{km}$.

The displacement from B to C is $-7\mathbf{i} + 12\mathbf{j}\,\mathrm{km}$.

a Find the magnitude of the displacement from A to C.

b Find the total distance the girl has cycled in getting from A to C.

c Work out the angle $\overrightarrow{AC}$ makes with the unit vector $\mathbf{i}$.

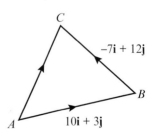

Mixed exercise 8

(P) 1 The motion of a cricket ball after it is hit until it lands on the cricket pitch can be modelled using the equation $h = \frac{1}{10}(24x - 3x^2)$, where h m is the vertical height of the ball above the cricket pitch and x m is the horizontal distance from where it was hit. Find:

> **Hint** The path of the cricket ball is modelled as a quadratic curve. Draw a sketch for the model and use the symmetry of the curve.

   a the vertical height of the ball when it is at a horizontal distance of 2 m from where it was hit

   b the two horizontal distances for which the height of the ball was 2.1 m.

   Given that the model is valid from when the ball is hit to when it lands on the cricket pitch:

   c find the values of x for which the model is valid

   d work out the maximum height of the cricket ball.

(P) 2 A diver dives from a diving board into a swimming pool with a depth of 4.5 m. The height of the diver above the water, h m, can be modelled using $h = 10 - 0.58x^2$ for $0 \leqslant x \leqslant 5$, where x m is the horizontal distance from the end of the diving board.

   a Find the height of the diver when $x = 2$ m.

   b Find the horizontal distance from the end of the diving board to the point where the diver enters the water.

   In this model the diver is modelled as particle.

   c Describe the effects of this modelling assumption.

   d Comment on the validity of this modelling assumption for the motion of the diver after she enters the water.

3 Make a list of the assumptions you might make to create simple models of the following:

   a The motion of a man skiing down a snow-covered slope.

   b The motion of a yo-yo on a string.

   In each case, describe the effects of the modelling assumptions.

4 Convert to SI units:

   a 2.5 km per minute   **b** 0.6 kg cm^{-2} **c** 1.2×10^3 g cm^{-3}

5 A man throws a bowling ball in a bowling alley.

   a Make a list of the assumptions you might make to create a simple model of the motion of the bowling ball.

   b Taking the direction that the ball travels in as the positive direction, state with a reason whether each of the following are likely to be positive or negative:

      i the velocity   **ii** the acceleration.

6 A train engine pulling a truck starts at station A then travels in a straight line to station B. It then moves back from station B to station A and on to station C as shown in the diagram.

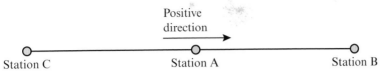

Positive direction

Station C   Station A   Station B

Hint The **sign** of something means whether it is positive or negative.

What is the sign of the velocity and displacement on the journey from:

a station A to station B **b** station B to station A **c** station A to station C?

7 The acceleration of a boat is given by $\mathbf{a} = -0.05\mathbf{i} + 0.15\mathbf{j}\,\text{m s}^{-2}$. Find:

  a the magnitude of the acceleration

  b the angle the direction of the acceleration vector makes with the unit vector $\mathbf{i}$.

8 The velocity of a toy car is given by $\mathbf{v} = 3.5\mathbf{i} - 2.5\mathbf{j}\,\text{m s}^{-1}$. Find:

  a the speed of the toy car

  b the angle the direction of motion of the toy car makes with the unit vector $\mathbf{j}$.

9 A plane flies from P to Q and then from Q to R.

The displacement from P to Q is $100\mathbf{i} + 80\mathbf{j}\,\text{m}$.

The displacement from Q to R is $50\mathbf{i} - 30\mathbf{j}\,\text{m}$.

  a Find the magnitude of the displacement from P to R.

  b Find the total distance the plane has travelled in getting from P to R.

  c Find the angle the vector $\overrightarrow{PQ}$ makes with the unit vector $\mathbf{j}$.

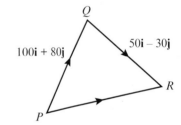

Summary of key points

1 Mathematical models can be constructed to simulate real-life situations.

2 Modelling assumptions can be used to simplify your calculations.

3 The base SI units most commonly used in mechanics are:

Quantity	Unit	Symbol
Mass	kilogram	kg
Length/displacement	metre	m
Time	second	s

4 A vector is a quantity which has both magnitude and direction.

5 A scalar quantity has magnitude only.

6 Distance is the magnitude of the displacement vector.

7 Speed is the magnitude of the velocity vector.

9 Constant acceleration

Objectives

After completing this chapter you should be able to:

● Understand and interpret displacement–time graphs → **pages 131–133**

● Understand and interpret velocity–time graphs → **pages 133–136**

● Derive the constant acceleration formulae and use them to solve problems → **pages 137–146**

● Use the constant acceleration formulae to solve problems involving vertical motion under gravity → **pages 146–152**

Prior knowledge check

1 For each graph find:
 i the gradient
 ii the shaded area under the graph.

a **b** **c**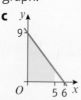

2 A car travels for 45 minutes at an average speed of 35 mph. Find the distance travelled.

3 a Solve the simultaneous equations:
$$3x - 2y = 9$$
$$x + 4y + 4 = 0$$

b Solve $2x^2 + 3x - 7 = 0$. Give your answers to 3 s.f.

A body falling freely under **gravity** can be modelled as having **constant acceleration**. You can use this to estimate the time it will take a cliff diver to reach the water.

→ **Exercise 9E Q1**

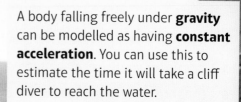

9.1 Displacement–time graphs

You can represent the motion of an object on a displacement–time graph. Displacement is always plotted on the vertical axis and time on the horizontal axis.

In these graphs s represents the displacement of an object from a given point in metres and t represents the time taken in seconds.

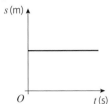

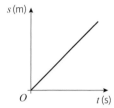

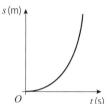

There is no change in the displacement over time and the object is stationary.

The displacement increases at a constant rate over time and the object is moving with constant velocity.

The displacement is increasing at a greater rate as time increases. The velocity is increasing and the object is accelerating.

- **Velocity is the rate of change of displacement.**
  - **On a displacement–time graph the gradient represents the velocity.**
  - **If the displacement–time graph is a straight line, then the velocity is constant.**

- **Average velocity $=\dfrac{\textbf{displacement from starting point}}{\textbf{time taken}}$**

- **Average speed $=\dfrac{\textbf{total distance travelled}}{\textbf{time taken}}$**

Example 1

A cyclist rides in a straight line for 20 minutes. She waits for half an hour, then returns in a straight line to her starting point in 15 minutes. This is a displacement–time graph for her journey.

a Work out the average velocity for each stage of the journey in km h⁻¹.

b Write down the average velocity for the whole journey.

c Work out the average speed for the whole journey.

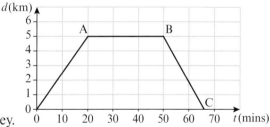

a Journey from O to A: time = 20 mins; displacement = 5 km

$\text{Average velocity} = \dfrac{5}{20} = 0.25\text{ km min}^{-1}$

$0.25 \times 60 = 15\text{ km h}^{-1}$

Journey from A to B: no change in displacement so average velocity = 0

Journey from B to C: time = 15 mins; displacement = −5 km

$\text{Average velocity} = \dfrac{-5}{15} = -\dfrac{1}{3}\text{ km min}^{-1}$

$-\dfrac{1}{3} \times 60 = -20\text{ km h}^{-1}$

To convert from km min⁻¹ to km h⁻¹ multiply by 60.

A horizontal line on the graph indicates the cyclist is stationary.

The cyclist starts with a displacement of 5 km and finishes with a displacement of 0 km, so the change in displacement is −5 km, and velocity will be negative.

131

b The displacement for the whole journey is 0 so average velocity is 0.

> At *C* the cyclist has returned to the starting point.

c Total time = 65 mins

Total distance travelled is 5 + 5 = 10 km

Average speed = $\dfrac{10}{65} = \dfrac{2}{13}$ km min⁻¹

$\dfrac{2}{13} \times 60 = 9.2$ km h⁻¹ (2 s.f.)

> The cyclist has travelled 5 km away from the starting point and then 5 km back to the starting point.

Exercise 9A

1 This is a displacement–time graph for a car travelling along a straight road. The journey is divided into 5 stages labelled *A* to *E*.

   a Work out the average velocity for each stage of the journey.

   b State the average velocity for the whole journey.

   c Work out the average speed for the whole journey.

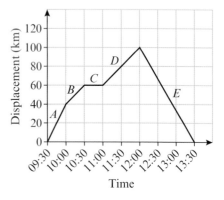

P **2** Khalid drives from his home to a hotel. He drives for $2\frac{1}{2}$ hours at an average velocity of 60 km h⁻¹. He then stops for lunch before continuing to his hotel. The diagram shows a displacement–time graph for Khalid's journey.

   a Work out the displacement of the hotel from Khalid's home.

   b Work out Khalid's average velocity for his whole journey.

Problem-solving

You need to work out the scale on the vertical axis.

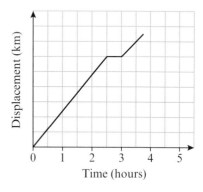

P **3** Sarah left home at 10:00 and cycled north in a straight line. The diagram shows a displacement–time graph for her journey.

   a Work out Sarah's velocity between 10:00 and 11:00.

On her return journey, Sarah continued past her home for 3 km before returning.

   b Estimate the time that Sarah passed her home.

   c Work out Sarah's velocity for each of the last two stages of her journey.

   d Calculate Sarah's average speed for her entire journey.

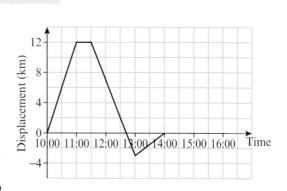

P **4** A ball is thrown vertically up in the air and falls to the ground. This is a displacement–time graph for the motion of the ball.

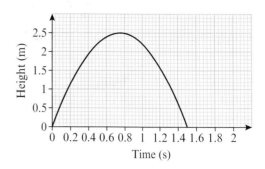

a Find the maximum height of the ball and the time at which it reaches that height.

b Write down the velocity of the ball when it reaches its highest point.

c Describe the motion of the ball:

   i from the time it is thrown to the time it reaches its highest point

   ii after reaching its highest point.

Hint To describe the motion you should state the direction of travel of the ball and whether it is accelerating or decelerating.

9.2 Velocity–time graphs

You can represent the motion of an object on a velocity–time graph. Velocity is always plotted on the vertical axis and time on the horizontal axis.

In these graphs v represents the velocity of an object in metres per second and t represents the time taken in seconds.

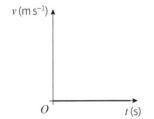

The velocity is zero and the object is stationary.

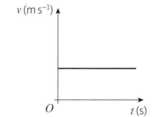

The velocity is unchanging and the object is moving with constant velocity.

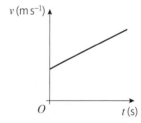

The velocity is increasing at a constant rate and the object is moving with constant acceleration.

- **Acceleration is the rate of change of velocity.**
  - **In a velocity–time graph the gradient represents the acceleration.**
  - **If the velocity–time graph is a straight line, then the acceleration is constant.**

Notation Negative acceleration is sometimes described as deceleration or retardation.

This velocity–time graph represents the motion of an object travelling in a straight line at constant velocity V m s⁻¹ for time T seconds.

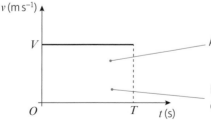

Area under the graph = $V \times T$

For an object with constant velocity, displacement = velocity × time

- **The area between a velocity–time graph and the horizontal axis represents the distance travelled.**
  - **For motion in a straight line with positive velocity, the area under the velocity–time graph up to a point t represents the displacement at time t.**

Example 2

The figure shows a velocity–time graph illustrating the motion of a cyclist moving along a straight road for a period of 12 seconds. For the first 8 seconds, she moves at a constant speed of $6\,\text{m s}^{-1}$. She then decelerates at a constant rate, stopping after a further 4 seconds.

a Find the displacement from the starting point of the cyclist after this 12 second period.

b Work out the rate at which the cyclist decelerates.

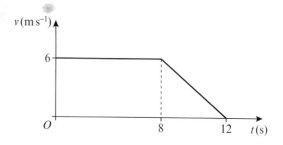

a The displacement s after 12 s is given by the area under the graph.

$$s = \frac{1}{2}(a + b)h$$

$$= \frac{1}{2}(8 + 12) \times 6$$

$$= 10 \times 6 = 60$$

The displacement of the cyclist after 12 s is 60 m.

Model the cyclist as a particle moving in a straight line.

The displacement is represented by the area of the trapezium with these sides.

You can use the formula for the area of a trapezium to calculate this area.

b The acceleration is the gradient of the slope.

$$a = \frac{-6}{4} = -1.5$$

The deceleration is $1.5\,\text{m s}^{-2}$.

The gradient is given by

$$\frac{\text{difference in the } v\text{-coordinates}}{\text{difference in the } t\text{-coordinates}}$$

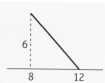

Example 3

A particle moves along a straight line. The particle accelerates uniformly from rest to a velocity of $8\,\text{m s}^{-1}$ in T seconds. The particle then travels at a constant velocity of $8\,\text{m s}^{-1}$ for $5T$ seconds. The particle then decelerates uniformly to rest in a further 40 s.

a Sketch a velocity–time graph to illustrate the motion of the particle.

Given that the total displacement of the particle is 600 m:

b find the value of T.

a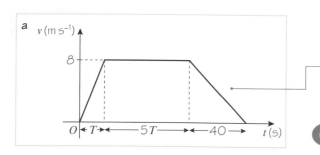

If the particle accelerates from rest and decelerates to rest this means the initial and final velocities are zero.

Online Explore how the area of the trapezium changes as the value of T changes using technology.

b The area of the trapezium is:

$s = \frac{1}{2}(a + b)h$

$= \frac{1}{2}(5T + 6T + 40) \times 8$

$= 4(11T + 40)$

The displacement is 600 m.

$4(11T + 40) = 600$

$44T + 160 = 600$

$T = \dfrac{600 - 160}{44} = 10$

The length of the shorter of the two parallel sides is $5T$. The length of the longer side is $T + 5T + 40 = 6T + 40$.

Problem-solving

The displacement is equal to the area of the trapezium. Write an equation and solve it to find T.

Exercise 9B

1 The diagram shows the velocity–time graph of the motion of an athlete running along a straight track. For the first 4 s, he accelerates uniformly from rest to a velocity of $9 \, \text{m s}^{-1}$.
This velocity is then maintained for a further 8 s. Find:

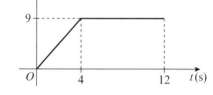

a the rate at which the athlete accelerates

b the displacement from the starting point of the athlete after 12 s.

2 A car is moving along a straight road. When $t = 0$ s, the car passes a point A with velocity $10 \, \text{m s}^{-1}$ and this velocity is maintained until $t = 30$ s. The driver then applies the brakes and the car decelerates uniformly, coming to rest at the point B when $t = 42$ s.

a Sketch a velocity–time graph to illustrate the motion of the car.

b Find the distance from A to B.

(E) 3 The diagram shows the velocity–time graph of the motion of a cyclist riding along a straight road. She accelerates uniformly from rest to $8 \, \text{m s}^{-1}$ in 20 s. She then travels at a constant velocity of $8 \, \text{m s}^{-1}$ for 40 s. She then decelerates uniformly to rest in 15 s. Find:

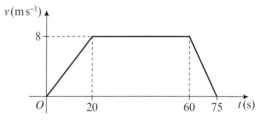

a the acceleration of the cyclist in the first 20 s of motion (**2 marks**)

b the deceleration of the cyclist in the last 15 s of motion (**2 marks**)

c the displacement from the starting point of the cyclist after 75 s. (**2 marks**)

(E) 4 A motorcyclist starts from rest at a point S on a straight race track. He moves with constant acceleration for 15 s, reaching a velocity of $30 \, \text{m s}^{-1}$. He then travels at a constant velocity of $30 \, \text{m s}^{-1}$ for T seconds. Finally he decelerates at a constant rate coming to rest at a point F, 25 s after he begins to decelerate.

a Sketch a velocity–time graph to illustrate the motion. (**3 marks**)

Given that the distance between S and F is 2.4 km:

b calculate the time the motorcyclist takes to travel from S to F. (**3 marks**)

Ⓔ **5** A train starts from a station X and moves with constant acceleration of $0.6\,\mathrm{m\,s^{-2}}$ for $20\,\mathrm{s}$. The velocity it has reached after $20\,\mathrm{s}$ is then maintained for T seconds. The train then decelerates from this velocity to rest in a further $40\,\mathrm{s}$, stopping at a station Y.

   a Sketch a velocity–time graph to illustrate the motion of the train. **(3 marks)**

   Given that the distance between the stations is $4.2\,\mathrm{km}$, find:

   b the value of T **(3 marks)**

   c the distance travelled by the train while it is moving with constant velocity. **(2 marks)**

Ⓔ **6** A particle moves along a straight line. The particle accelerates from rest to a velocity of $10\,\mathrm{m\,s^{-1}}$ in $15\,\mathrm{s}$. The particle then moves at a constant velocity of $10\,\mathrm{m\,s^{-1}}$ for a period of time. The particle then decelerates uniformly to rest. The period of time for which the particle is travelling at a constant velocity is 4 times the period of time for which it is decelerating.

   a Sketch a velocity–time graph to illustrate the motion of the particle. **(3 marks)**

   Given that the displacement from the starting point of the particle after it comes to rest is $480\,\mathrm{m}$

   b find the total time for which the particle is moving. **(3 marks)**

Ⓔ **7** A particle moves $100\,\mathrm{m}$ in a straight line. The diagram is a sketch of a velocity–time graph of the motion of the particle. The particle starts with velocity $u\,\mathrm{m\,s^{-1}}$ and accelerates to a velocity of $10\,\mathrm{m\,s^{-1}}$ in $3\,\mathrm{s}$. The velocity of $10\,\mathrm{m\,s^{-1}}$ is maintained for $7\,\mathrm{s}$ and then the particle decelerates to rest in a further $2\,\mathrm{s}$. Find:

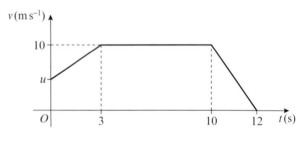

   a the value of u **(3 marks)**

   b the acceleration of the particle in the first $3\,\mathrm{s}$ of motion. **(3 marks)**

Ⓔ **8** A motorcyclist M leaves a road junction at time $t = 0\,\mathrm{s}$. She accelerates from rest at a rate of $3\,\mathrm{m\,s^{-2}}$ for $8\,\mathrm{s}$ and then maintains the velocity she has reached. A car C leaves the same road junction as M at time $t = 0\,\mathrm{s}$. The car accelerates from rest to $30\,\mathrm{m\,s^{-1}}$ in $20\,\mathrm{s}$ and then maintains the velocity of $30\,\mathrm{m\,s^{-1}}$. C passes M as they both pass a pedestrian.

   a On the same diagram, sketch velocity–time graphs to illustrate the motion of M and C. **(3 marks)**

   b Find the distance of the pedestrian from the road junction. **(3 marks)**

Challenge

The graph shows the velocity of an object travelling in a straight line during a 10-second time interval.

a After how long did the object change direction?

b Work out the total distance travelled by the object.

c Work out the displacement from the starting point of the object after:

   i 6 seconds   **ii** 10 seconds.

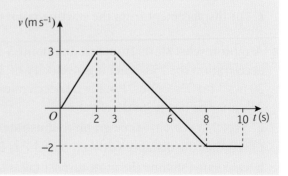

9.3 Constant acceleration formulae 1

A standard set of letters is used for the motion of an object moving in a straight line with constant acceleration.

- s is the displacement.
- u is the initial velocity.
- v is the final velocity.
- a is the acceleration.
- t is the time.

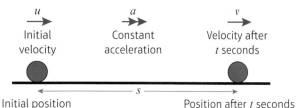

You can use these letters to label a velocity–time graph representing the motion of a particle moving in a straight line accelerating from velocity u at time 0 to velocity v at time t.

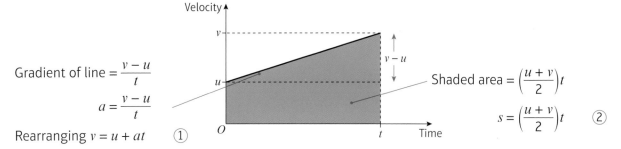

Gradient of line $= \dfrac{v - u}{t}$

$a = \dfrac{v - u}{t}$

Rearranging $v = u + at$ ①

Shaded area $= \left(\dfrac{u + v}{2}\right)t$

$s = \left(\dfrac{u + v}{2}\right)t$ ②

- $v = u + at$ ①
- $s = \left(\dfrac{u + v}{2}\right)t$ ②

You need to know how to derive these formulae from the velocity–time graph.

Hint   Formula ① does not involve s and formula ② does not involve a.

Links   These formulae can also be derived using calculus.   → **Chapter 11**

Example   4

A cyclist is travelling along a straight road. She accelerates at a constant rate from a velocity of $4\,\mathrm{m\,s^{-1}}$ to a velocity of $7.5\,\mathrm{m\,s^{-1}}$ in 40 seconds. Find:

a the distance she travels in these 40 seconds

b her acceleration in these 40 seconds.

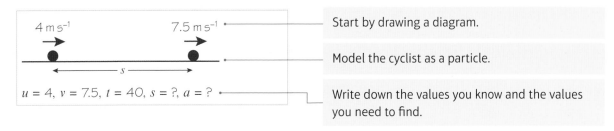

Start by drawing a diagram.

Model the cyclist as a particle.

Write down the values you know and the values you need to find.

a $s = \left(\dfrac{u + v}{2}\right)t$

$= \left(\dfrac{4 + 7.5}{2}\right) \times 40$

$= 230$

The distance the cyclist travels is 230 m.

You need a and you know v, u and t so you can use $v = u + at$.

Substitute the values you know into the formula. You can solve this equation to find a.

b $v = u + at$

$7.5 = 4 + 40a$

$a = \dfrac{7.5 - 4}{40} = 0.0875$

The acceleration of the cyclist is 0.0875 m s^{-2}.

You could rearrange the formula before you substitute the values:

$a = \dfrac{v - u}{t}$

In real-life situations values for the acceleration are often quite small.
Large accelerations feel unpleasant and may be dangerous.

Example 5

A particle moves in a straight line from a point A to a point B with constant deceleration 1.5 m s^{-2}. The velocity of the particle at A is 8 m s^{-1} and the velocity of the particle at B is 2 m s^{-1}. Find:

a the time taken for the particle to move from A to B

b the distance from A to B.

After reaching B the particle continues to move along the straight line with constant deceleration 1.5 m s^{-2}. The particle is at the point C 6 seconds after passing through the point A. Find:

c the velocity of the particle at C

d the distance from A to C.

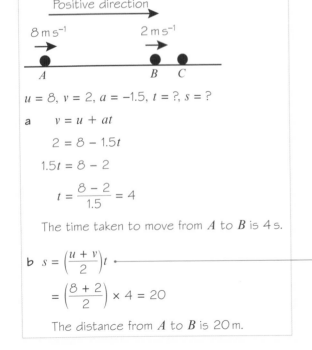

$u = 8,\ v = 2,\ a = -1.5,\ t = ?,\ s = ?$

a $v = u + at$

$2 = 8 - 1.5t$

$1.5t = 8 - 2$

$t = \dfrac{8 - 2}{1.5} = 4$

The time taken to move from A to B is 4 s.

b $s = \left(\dfrac{u + v}{2}\right)t$

$= \left(\dfrac{8 + 2}{2}\right) \times 4 = 20$

The distance from A to B is 20 m.

Problem-solving

It's always a good idea to draw a sketch showing the positions of the particle. Mark the positive direction on your sketch, and remember that when the particle is **decelerating**, your value of a will be **negative**.

You can use your answer from part **a** as the value of t.

c $u = 8$, $a = -15$, $t = 6$, $v = ?$

$v = u + at$

$= 8 + (-1.5) \times 6$

$= 8 - 9 = -1$

The velocity at C is negative. This means that the particle is moving from right to left.

The velocity of the particle is $1\,\text{m s}^{-1}$ in the direction $\overrightarrow{BA}$.

Remember that to specify a velocity it is necessary to give speed and direction.

d $s = \left(\dfrac{u + v}{2}\right)t$

$= \left(\dfrac{8 + (-1)}{2}\right) \times 6$

Make sure you use the correct sign when substituting a negative value into a formula.

The distance from A to C is $21\,\text{m}$.

Convert all your measurements into base SI units before substituting values into the formulae.

Example 6

A car moves from traffic lights along a straight road with constant acceleration. The car starts from rest at the traffic lights and 30 seconds later the car passes a speed-trap where it is registered as travelling at $45\,\text{km h}^{-1}$. Find:

a the acceleration of the car   **b** the distance between the traffic lights and the speed-trap.

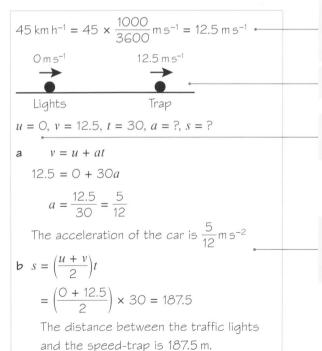

$45\,\text{km h}^{-1} = 45 \times \dfrac{1000}{3600}\,\text{m s}^{-1} = 12.5\,\text{m s}^{-1}$

Convert into SI units, using:
$1\,\text{km} = 1000\,\text{m}$
$1\,\text{hour} = 60 \times 60\,\text{s} = 3600\,\text{s}$

$0\,\text{m s}^{-1}$ $12.5\,\text{m s}^{-1}$

Lights   Trap

Model the car as a particle and draw a diagram.

$u = 0$, $v = 12.5$, $t = 30$, $a = ?$, $s = ?$

The car starts from rest, so the initial velocity is zero.

a $v = u + at$

$12.5 = 0 + 30a$

$a = \dfrac{12.5}{30} = \dfrac{5}{12}$

The acceleration of the car is $\dfrac{5}{12}\,\text{m s}^{-2}$

This is an exact answer. If you want to give an answer using decimals, you should round to three significant figures.

b $s = \left(\dfrac{u + v}{2}\right)t$

$= \left(\dfrac{0 + 12.5}{2}\right) \times 30 = 187.5$

The distance between the traffic lights and the speed-trap is $187.5\,\text{m}$.

Exercise 9C

1 A particle is moving in a straight line with constant acceleration $3 \, \text{m s}^{-2}$. At time $t = 0$, the velocity of the particle is $2 \, \text{m s}^{-1}$. Find the velocity of the particle at time $t = 6 \, \text{s}$.

2 A car is approaching traffic lights. The car is travelling with velocity $10 \, \text{m s}^{-1}$. The driver applies the brakes to the car and the car comes to rest with constant deceleration in $16 \, \text{s}$. Modelling the car as a particle, find the deceleration of the car.

3 A car accelerates uniformly while travelling on a straight road. The car passes two signposts $360 \, \text{m}$ apart. The car takes $15 \, \text{s}$ to travel from one signpost to the other. When passing the second signpost, it has velocity $28 \, \text{m s}^{-1}$. Find the velocity of the car at the first signpost.

4 A cyclist is moving along a straight road from A to B with constant acceleration $0.5 \, \text{m s}^{-2}$. Her velocity at A is $3 \, \text{m s}^{-1}$ and it takes her 12 seconds to cycle from A to B. Find:
   a her velocity at B
   b the distance from A to B.

5 A particle is moving along a straight line with constant acceleration from a point A to a point B, where $AB = 24 \, \text{m}$. The particle takes $6 \, \text{s}$ to move from A to B and the velocity of the particle at B is $5 \, \text{m s}^{-1}$. Find:
   a the velocity of the particle at A
   b the acceleration of the particle.

6 A particle moves in a straight line from a point A to a point B with constant deceleration $1.2 \, \text{m s}^{-2}$. The particle takes $6 \, \text{s}$ to move from A to B. The speed of the particle at B is $2 \, \text{m s}^{-1}$ and the direction of motion of the particle has not changed. Find:
   a the speed of the particle at A
   b the distance from A to B.

(P) 7 A train, travelling on a straight track, is slowing down with constant deceleration $0.6 \, \text{m s}^{-2}$. The train passes one signal with speed $72 \, \text{km h}^{-1}$ and a second signal $25 \, \text{s}$ later. Find:

> **Hint** Convert the speeds into m s^{-1} before substituting.

   a the velocity, in km h^{-1}, of the train as it passes the second signal
   b the distance between the signals.

8 A particle moves in a straight line from a point A to a point B with a constant deceleration of $4 \, \text{m s}^{-2}$. At A the particle has velocity $32 \, \text{m s}^{-1}$ and the particle comes to rest at B. Find:
   a the time taken for the particle to travel from A to B
   b the distance between A and B.

(E) 9 A skier travelling in a straight line up a hill experiences a constant deceleration. At the bottom of the hill, the skier has a velocity of $16 \, \text{m s}^{-1}$ and, after moving up the hill for $40 \, \text{s}$, he comes to rest. Find:
   a the deceleration of the skier   **(2 marks)**
   b the distance from the bottom of the hill to the point where the skier comes to rest. **(4 marks)**

(E) **10** A particle is moving in a straight line with constant acceleration. The points A, B and C lie on this line. The particle moves from A through B to C. The velocity of the particle at A is $2\,\text{m s}^{-1}$ and the velocity of the particle at B is $7\,\text{m s}^{-1}$. The particle takes $20\,\text{s}$ to move from A to B.

    a Find the acceleration of the particle. **(2 marks)**

    The velocity of the particle is C is $11\,\text{m s}^{-1}$. Find:

    b the time taken for the particle to move from B to C **(2 marks)**

    c the distance between A and C. **(3 marks)**

(E) **11** A particle moves in a straight line from A to B with constant acceleration $1.5\,\text{m s}^{-2}$. It then moves along the same straight line from B to C with a different acceleration. The velocity of the particle at A is $1\,\text{m s}^{-1}$ and the velocity of the particle at C is $43\,\text{m s}^{-1}$. The particle takes $12\,\text{s}$ to move from A to B and $10\,\text{s}$ to move from B to C. Find:

    a the velocity of the particle at B **(2 marks)**

    b the acceleration of the particle as it moves from B to C **(2 marks)**

    c the distance from A to C. **(3 marks)**

(E/P) **12** A cyclist travels with constant acceleration $x\,\text{m s}^{-2}$, in a straight line, from rest to $5\,\text{m s}^{-1}$ in $20\,\text{s}$. She then decelerates from $5\,\text{m s}^{-1}$ to rest with constant deceleration $\frac{1}{2}x\,\text{m s}^{-2}$. Find:

    a the value of x **(2 marks)**

    b the total distance she travelled. **(4 marks)**

> **Problem-solving**
>
> You could sketch a velocity–time graph of the cyclist's motion and use the area under the graph to find the total distance travelled.

(E/P) **13** A particle is moving with constant acceleration in a straight line. It passes through three points, A, B and C, with velocities $20\,\text{m s}^{-1}$, $30\,\text{m s}^{-1}$ and $45\,\text{m s}^{-1}$ respectively. The time taken to move from A to B is t_1 seconds and the time taken to move from B to C is t_2 seconds.

    a Show that $\dfrac{t_1}{t_2} = \dfrac{2}{3}$. **(3 marks)**

    Given also that the total time taken for the particle to move from A to C is $50\,\text{s}$:

    b find the distance between A and B. **(5 marks)**

> **Challenge**
>
> A particle moves in a straight line from A to B with constant acceleration. The particle moves from A with velocity $3\,\text{m s}^{-1}$. It reaches point B with velocity $5\,\text{m s}^{-1}$ t seconds later.
>
> One second after the first particle leaves point A, a second particle also starts to move in a straight line from A to B with constant acceleration. Its velocity at point A is $4\,\text{m s}^{-1}$ and it reaches point B with velocity $8\,\text{m s}^{-1}$ at the same time as the first particle.
>
> Find:
>
> **a** the value of t
>
> **b** the distance between A and B.
>
> > **Problem-solving**
> >
> > The time taken for the second particle to travel from A to B is $(t - 1)$ seconds.

9.4 Constant acceleration formulae 2

You can use the formulae $v = u + at$ and $s = \left(\dfrac{u + v}{2}\right)t$ to work out three more formulae.

You can eliminate t from the formulae for constant acceleration.

$$t = \frac{v - u}{a}$$
— Rearrange the formula $v = u + at$ to make t the subject.

$$s = \left(\frac{u + v}{2}\right)\left(\frac{v - u}{a}\right)$$
— Substitute this expression for t into $s = \left(\dfrac{u + v}{2}\right)t$.

$$2as = v^2 - u^2$$

- $v^2 = u^2 + 2as$ —— Multiply out the brackets and rearrange.

You can also eliminate v from the formulae for constant acceleration.

$$s = \left(\frac{u + u + at}{2}\right)t$$
— Substitute $v = u + at$ into $s = \left(\dfrac{u + v}{2}\right)t$.

$$s = \left(\frac{2u}{2} + \frac{at}{2}\right)t$$

$$s = \left(u + \frac{1}{2}at\right)t$$
— Multiply out the brackets and rearrange.

- $s = ut + \dfrac{1}{2}at^2$

Finally, you can eliminate u by substituting into this formula:

$$s = (v - at)t + \frac{1}{2}at^2$$
— Substitute $u = v - at$ into $s = ut + \dfrac{1}{2}at^2$.

- $s = vt - \dfrac{1}{2}at^2$

■ **You need to be able to use and to derive the five formulae for solving problems about particles moving in a straight line with constant acceleration.**

- $v = u + at$

- $s = \left(\dfrac{u + v}{2}\right)t$

- $v^2 = u^2 + 2as$

- $s = ut + \dfrac{1}{2}at^2$

- $s = vt - \dfrac{1}{2}at^2$

Watch out These five formulae are sometimes referred to as the **kinematics formulae** or *suvat formulae*. They are given in the formulae booklet.

Example 7

A particle is moving along a straight line from A to B with constant acceleration $5\,\text{m s}^{-2}$.
The velocity of the particle at A is $3\,\text{m s}^{-1}$ in the direction $\overrightarrow{AB}$. The velocity of the particle at B is $18\,\text{m s}^{-1}$ in the same direction. Find the distance from A to B.

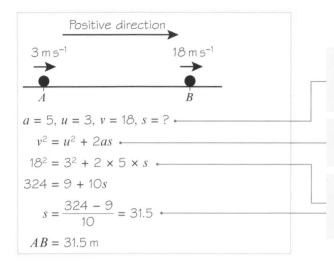

$a = 5,\ u = 3,\ v = 18,\ s = ?$

$v^2 = u^2 + 2as$

$18^2 = 3^2 + 2 \times 5 \times s$

$324 = 9 + 10s$

$s = \dfrac{324 - 9}{10} = 31.5$

$AB = 31.5\,\text{m}$

Write down the values you know and the values you need to find. This will help you choose the correct formula.

t is not involved so choose the formula that does not have t in it.

Substitute in the values you are given and solve the equation for s. This gives the distance you were asked to find.

Example 8

A particle is moving in a straight horizontal line with constant deceleration $4\,\text{m s}^{-2}$. At time $t = 0$ the particle passes through a point O with speed $13\,\text{m s}^{-1}$ travelling towards a point A, where $OA = 20\,\text{m}$. Find:

a the times when the particle passes through A

b the value of t when the particle returns to O.

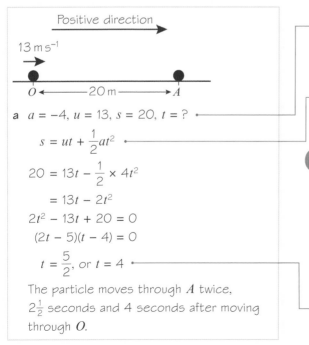

a $a = -4,\ u = 13,\ s = 20,\ t = ?$

$s = ut + \dfrac{1}{2}at^2$

$20 = 13t - \dfrac{1}{2} \times 4t^2$

$\quad = 13t - 2t^2$

$2t^2 - 13t + 20 = 0$

$(2t - 5)(t - 4) = 0$

$t = \dfrac{5}{2},\text{ or } t = 4$

The particle moves through A twice, $2\frac{1}{2}$ seconds and 4 seconds after moving through O.

The particle is decelerating so the value of a is negative.

You are told the values of a, u and s and asked to find t. You are given no information about v and are not asked to find it so you choose the formula without v.

Problem-solving

When you use $s = ut + \frac{1}{2}at^2$ with an unknown value of t you obtain a quadratic equation in t. You can solve this equation by factorising, or using the quadratic formula, $x = \dfrac{-b \pm \sqrt{b^2 - 4ac}}{2a}$.

There are two answers. Both are correct. The particle moves from O to A, goes beyond A and then turns round and returns to A.

b The particle returns to O when $s = 0$.

$s = 0$, $u = 13$, $a = -4$, $t = ?$

$s = ut + \dfrac{1}{2}at^2$

$0 = 13t - 2t^2$

$\quad = t(13 - 2t)$

$t = 0$, or $t = \dfrac{13}{2}$

The particle returns to O 6.5 seconds after it first passed through O.

> When the particle returns to O, its displacement (distance) from O is zero.

> The first solution ($t = 0$) represents the starting position of the particle. The other solution ($t = \frac{13}{2}$) tells you when the particle returns to O.

Example 9

A particle P is moving on the x-axis with constant deceleration $2.5\,\text{m s}^{-2}$. At time $t = 0$, the particle P passes through the origin O, moving in the positive direction of x with speed $15\,\text{m s}^{-1}$. Find:

a the time between the instant when P first passes through O and the instant when it returns to O

b the total distance travelled by P during this time.

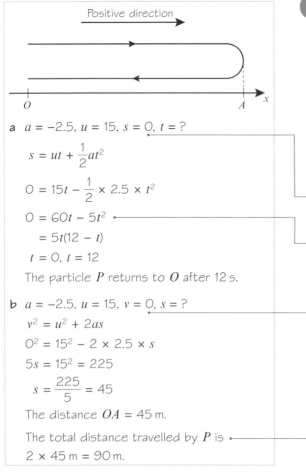

Positive direction

O $\qquad\qquad\qquad\qquad A \quad x$

a $a = -2.5$, $u = 15$, $s = 0$, $t = ?$

$s = ut + \dfrac{1}{2}at^2$

$0 = 15t - \dfrac{1}{2} \times 2.5 \times t^2$

$0 = 60t - 5t^2$

$\quad = 5t(12 - t)$

$t = 0$, $t = 12$

The particle P returns to O after 12 s.

b $a = -2.5$, $u = 15$, $v = 0$, $s = ?$

$v^2 = u^2 + 2as$

$0^2 = 15^2 - 2 \times 2.5 \times s$

$5s = 15^2 = 225$

$s = \dfrac{225}{5} = 45$

The distance $OA = 45$ m.

The total distance travelled by P is

2×45 m $= 90$ m.

> **Problem-solving**
>
> Before you start, draw a sketch so you can see what is happening. The particle moves through O with a positive velocity. As it is decelerating it slows down and will eventually have zero velocity at a point A, which you don't yet know. As the particle is still decelerating, its velocity becomes negative, so the particle changes direction and returns to O.

> When the particle returns to O, its displacement (distance) from O is zero.

> Multiply by 4 to get whole-number coefficients.

> At the furthest point from O, labelled A in the diagram, the particle changes direction. At that point, for an instant, the particle has zero velocity.

> In the 12 s the particle has been moving it has travelled to A and back. The total distance travelled is twice the distance OA.

Exercise 9D

1 A particle is moving in a straight line with constant acceleration $2.5\,\mathrm{m\,s^{-2}}$. It passes a point A with velocity $3\,\mathrm{m\,s^{-1}}$ and later passes through a point B, where $AB = 8\,\mathrm{m}$. Find the velocity of the particle as it passes through B.

2 A car is accelerating at a constant rate along a straight horizontal road. Travelling at $8\,\mathrm{m\,s^{-1}}$, it passes a pillar box and $6\,\mathrm{s}$ later it passes a sign. The distance between the pillar box and the sign is $60\,\mathrm{m}$. Find the acceleration of the car.

3 A cyclist travelling at $12\,\mathrm{m\,s^{-1}}$ applies her brakes and comes to rest after travelling $36\,\mathrm{m}$ in a straight line. Assuming that the brakes cause the cyclist to decelerate uniformly, find the deceleration.

4 A train is moving along a straight horizontal track with constant acceleration. The train passes a signal with a velocity of $54\,\mathrm{km\,h^{-1}}$ and a second signal with a velocity of $72\,\mathrm{km\,h^{-1}}$. The distance between the two signals is $500\,\mathrm{m}$. Find, in $\mathrm{m\,s^{-2}}$, the acceleration of the train.

5 A particle moves along a straight line, with constant acceleration, from a point A to a point B where $AB = 48\,\mathrm{m}$. At A the particle has velocity $4\,\mathrm{m\,s^{-1}}$ and at B it has velocity $16\,\mathrm{m\,s^{-1}}$. Find:
a the acceleration of the particle
b the time the particle takes to move from A to B.

6 A particle moves along a straight line with constant acceleration $3\,\mathrm{m\,s^{-2}}$. The particle moves $38\,\mathrm{m}$ in $4\,\mathrm{s}$. Find:
a the initial velocity of the particle
b the final velocity of the particle.

7 The driver of a car is travelling at $18\,\mathrm{m\,s^{-1}}$ along a straight road when she sees an obstruction ahead. She applies the brakes and the brakes cause the car to slow down to rest with a constant deceleration of $3\,\mathrm{m\,s^{-2}}$. Find:
a the distance travelled as the car decelerates
b the time it takes for the car to decelerate from $18\,\mathrm{m\,s^{-1}}$ to rest.

8 A stone is sliding across a frozen lake in a straight line. The initial speed of the stone is $12\,\mathrm{m\,s^{-1}}$. The friction between the stone and the ice causes the stone to slow down at a constant rate of $0.8\,\mathrm{m\,s^{-2}}$. Find:
a the distance moved by the stone before coming to rest
b the speed of the stone at the instant when it has travelled half of this distance.

9 A particle is moving along a straight line OA with constant acceleration $2.5\,\mathrm{m\,s^{-2}}$. At time $t = 0$, the particle passes through O with speed $8\,\mathrm{m\,s^{-1}}$ and is moving in the direction OA. The distance OA is $40\,\mathrm{m}$. Find:
a the time taken for the particle to move from O to A
b the speed of the particle at A. Give your answers to one decimal place.

10 A particle travels with uniform deceleration $2\,\mathrm{m\,s^{-2}}$ in a horizontal line. The points A and B lie on the line and $AB = 32\,\mathrm{m}$. At time $t = 0$, the particle passes through A with velocity $12\,\mathrm{m\,s^{-1}}$ in the direction $\overrightarrow{AB}$. Find:
a the values of t when the particle is at B
b the velocity of the particle for each of these values of t.

E/P **11** A particle is moving along the x-axis with constant deceleration 5 m s^{-2}. At time $t = 0$, the particle passes through the origin O with velocity 12 m s^{-1} in the positive direction. At time t seconds the particle passes through the point A with x-coordinate 8. Find:

> **Problem-solving**
>
> The particle will pass through A twice. Use $s = ut + \frac{1}{2}at^2$ to set up and solve a quadratic equation.

    a the values of t **(3 marks)**

    b the velocity of the particle as it passes through the point with x-coordinate -8.   **(3 marks)**

E **12** A particle P is moving on the x-axis with constant deceleration 4 m s^{-2}. At time $t = 0$, P passes through the origin O with velocity 14 m s^{-1} in the positive direction. The point A lies on the axis and $OA = 22.5 \text{ m}$. Find:

    a the difference between the times when P passes through A **(4 marks)**

    b the total distance travelled by P during the interval between these times.   **(3 marks)**

E/P **13** A car is travelling along a straight horizontal road with constant acceleration. The car passes over three consecutive points A, B and C where $AB = 100 \text{ m}$ and $BC = 300 \text{ m}$. The speed of the car at B is 14 m s^{-1} and the speed of the car at C is 20 m s^{-1}. Find:

    a the acceleration of the car   **(3 marks)**

    b the time take for the car to travel from A to C.   **(3 marks)**

E/P **14** Two particles P and Q are moving along the same straight horizontal line with constant accelerations 2 m s^{-2} and 3.6 m s^{-2} respectively. At time $t = 0$, P passes through a point A with speed 4 m s^{-1}. One second later Q passes through A with speed 3 m s^{-1}, moving in the same direction as P.

    a Write down expressions for the displacements of P and Q from A, in terms of t, where t seconds is the time after P has passed through A.   **(2 marks)**

    b Find the value of t where the particles meet.   **(3 marks)**

> **Problem-solving**
>
> When P and Q meet, their displacements from A are equal.

    c Find the distance of A from the point where the particles meet.   **(3 marks)**

E/P **15** In an orienteering competition, a competitor moves in a straight line past three checkpoints, P, Q and R, where $PQ = 2.4 \text{ km}$ and $QR = 11.5 \text{ km}$. The competitor is modelled as a particle moving with constant acceleration. She takes 1 hour to travel from P to Q and 1.5 hours to travel from Q to R. Find:

    a the acceleration of the competitor

    b her speed at the instant she passes P.   **(7 marks)**

9.5 Vertical motion under gravity

You can use the formulae for constant acceleration to model an object moving vertically under gravity.

- **The force of gravity causes all objects to accelerate towards the earth. If you ignore the effects of air resistance, this acceleration is constant. It does not depend on the mass of the object.**

As the force of gravity does not depend on mass, this means that in a vacuum an apple and a feather would both accelerate downwards at the same rate.

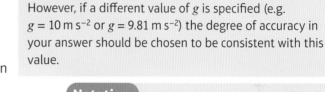

$g = 9.8\,\text{m s}^{-2}$ $g = 9.8\,\text{m s}^{-2}$

On earth, the acceleration due to gravity is represented by the letter g and is approximately $9.8\,\text{m s}^{-2}$.

The actual value of the acceleration can vary by very small amounts in different places due to the changing radius of the earth and height above sea level.

- **An object moving vertically under gravity can be modelled as a particle with a constant downward acceleration of $g = 9.8\,\text{m s}^{-2}$.**

Watch out In mechanics questions you will always use $g = 9.8\,\text{m s}^{-2}$ unless a question specifies otherwise. However, if a different value of g is specified (e.g. $g = 10\,\text{m s}^{-2}$ or $g = 9.81\,\text{m s}^{-2}$) the degree of accuracy in your answer should be chosen to be consistent with this value.

When solving problems about vertical motion you can choose the positive direction to be either upwards or downwards. Acceleration due to gravity is always downwards, so if the positive direction is upwards then $g = -9.8\,\text{m s}^{-2}$.

Notation The total time that an object is in motion from the time it is projected (thrown) upwards to the time it hits the ground is called the **time of flight**. The initial speed is sometimes called the **speed of projection**.

Example 10

A book falls off the top shelf of a bookcase. The shelf is $1.4\,\text{m}$ above a wooden floor. Find:
a the time the book takes to reach the floor, **b** the speed with which the book strikes the floor.

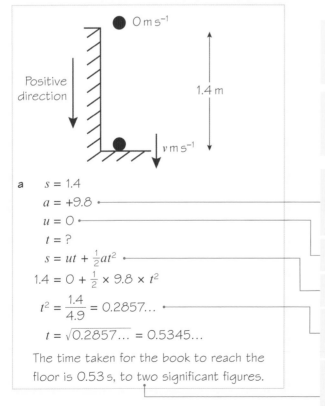

Model the book as a particle moving in a straight line with a constant acceleration of magnitude $9.8\,\text{m s}^{-2}$.

As the book is moving downwards throughout its motion, it is sensible to take the downwards direction as positive.

a $s = 1.4$
   $a = +9.8$
   $u = 0$
   $t = ?$
   $s = ut + \frac{1}{2}at^2$
   $1.4 = 0 + \frac{1}{2} \times 9.8 \times t^2$
   $t^2 = \dfrac{1.4}{4.9} = 0.2857\ldots$
   $t = \sqrt{0.2857\ldots} = 0.5345\ldots$

The time taken for the book to reach the floor is $0.53\,\text{s}$, to two significant figures.

You have taken the downwards direction as positive and gravity acts downwards. Here the acceleration is positive.

Assume the book has an initial speed of zero.

Choose the formula without v.

Solve the equation for t^2 and use your calculator to find the positive square root.

Give the answer to two significant figures to be consistent with the degree of accuracy used for the value of g.

b $s = 1.4$
   $a = 9.8$
   $u = 0$
   $v = ?$
   $v^2 = u^2 + 2as$ •————————— Choose the formula without t.

   $= 0^2 + 2 \times 9.8 \times 1.4 = 27.44$
   $v = \sqrt{27.44} = 5.238... \approx 5.2$
   The book hits the floor with speed
   $5.2 \, \text{m s}^{-1}$, to two significant figures. •———

Use unrounded values in your calculations, but give your final answer correct to two significant figures.

Example 11

A ball is projected vertically upwards, from a point X which is 7 m above the ground, with speed $21 \, \text{m s}^{-1}$. Find: **a** the greatest height above the ground reached by the ball, **b** the time of flight of the ball.

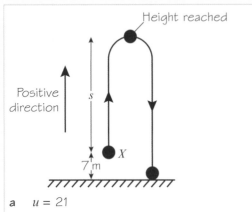

Problem-solving

In this sketch the upward and downwards motion have been sketched side by side. In reality they would be on top of one another, but drawing them separately makes it easier to see what is going on. Remember that X is 7 m above the ground, so mark this height on your sketch.

a $u = 21$
   $v = 0$ •
   $a = -9.8$
   $s = ?$
   $v^2 = u^2 + 2as$
   $0^2 = 21^2 + 2 \times (-9.8) \times s = 441 - 19.6s$
   $s = \dfrac{441}{19.6} = 22.5$ •
   $(22.5 + 7)\,\text{m} = 29.5\,\text{m}$ •
   Greatest height is 30 m (2 s.f.)

At its highest point, the ball is turning round. For an instant, it is neither going up nor down, so its speed is zero.

22.5 m is the distance the ball has moved above X but X is 7 m above the ground. You must add on another 7 m to get the greatest height above the ground reached by the ball.

b $s = -7$ •
   $u = 21$
   $a = -9.8$
   $t = ?$
   $s = ut + \frac{1}{2}at^2$
   $-7 = 21t - 4.9t^2$
   $4.9t^2 - 21t - 7 = 0$

The time of flight is the total time that the ball is in motion from the time that it is projected to the time that it stops moving. Here the ball will stop when it hits the ground. The point where the ball hits the ground is 7 m below the point from which it was projected so $s = -7$.

Online Use your calculator to check solutions to quadratic equations quickly.

$$t = \frac{-b \pm \sqrt{(b^2 - 4ac)}}{2a}$$

Rearrange the equation and use the quadratic formula.

$$= \frac{-(-21) \pm \sqrt{((-21)^2 - 4 \times 4.9 \times (-7))}}{2 \times 4.9}$$

$$= \frac{21 \pm \sqrt{578.2}}{9.8} \approx \frac{21 \pm 24.046}{9.8}$$

$$t \approx 4.5965,$$

$$\text{or } t \approx -0.3108$$

Take the positive answer and round to two significant figures.

Time of flight is 4.6 s (2 s.f.)

Example 12

A particle is projected vertically upwards from a point O with speed u m s^{-1}. The greatest height reached by the particle is 62.5 m above O. Find: **a** the value of u, **b** the total time for which the particle is 50 m or more above O.

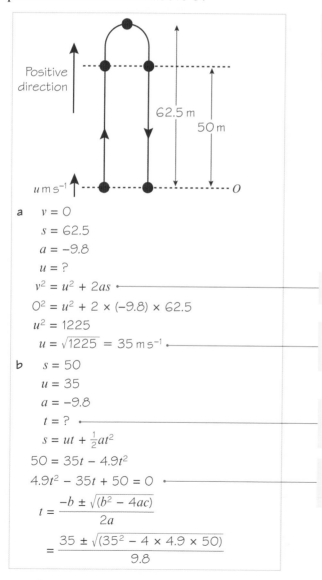

The particle will pass through the point 50 m above O twice: once on the way up and once on the way down.

a $v = 0$

$s = 62.5$

$a = -9.8$

$u = ?$

$v^2 = u^2 + 2as$

There is no t, so you choose the formula without t.

$0^2 = u^2 + 2 \times (-9.8) \times 62.5$

$u^2 = 1225$

$u = \sqrt{1225} = 35 \text{ m s}^{-1}$

In this part, you obtain an exact answer, so there is no need for approximation.

b $s = 50$

$u = 35$

$a = -9.8$

$t = ?$

Two values of t need to be found: one on the way up and one on the way down.

$s = ut + \frac{1}{2}at^2$

$50 = 35t - 4.9t^2$

$4.9t^2 - 35t + 50 = 0$

Write the equation in the form $ax^2 + bx + c = 0$ and use the quadratic formula.

$$t = \frac{-b \pm \sqrt{(b^2 - 4ac)}}{2a}$$

$$= \frac{35 \pm \sqrt{(35^2 - 4 \times 4.9 \times 50)}}{9.8}$$

$$= \frac{35 \pm \sqrt{245}}{9.8} \approx \frac{35 \pm 15.6525}{9.8}$$

$t \approx 5.1686\ldots$, or $t \approx 1.9742\ldots$

$(5.1686\ldots) - (1.9742\ldots) \approx 3.194$

Particle is 50 m or more above O for 3.2 s
(2 s.f.)

Between these two times the particle is always more than 50 m above O. You find the total time for which the particle is 50 m or more above O by finding the difference of these two values.

Example 13

A ball A falls vertically from rest from the top of a tower 63 m high. At the same time as A begins to fall, another ball B is projected vertically upwards from the bottom of the tower with speed 21 m s^{-1}. The balls collide. Find the distance of the point where the balls collide from the bottom of the tower.

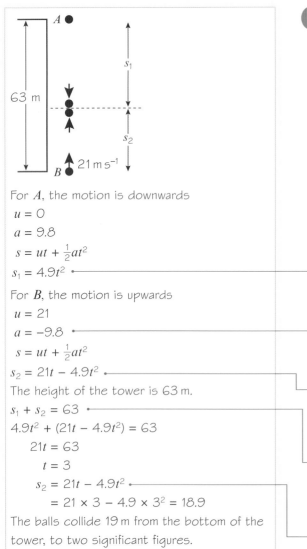

Problem-solving

You must take special care with problems where objects are moving in different directions. Here A is moving downwards and you will take the acceleration due to gravity as positive. However, B is moving upwards so for B the acceleration due to gravity is negative.

For A, the motion is downwards

$u = 0$

$a = 9.8$

$s = ut + \frac{1}{2}at^2$

$s_1 = 4.9t^2$

You cannot find s_1 at this stage. You have to express it in terms of t.

For B, the motion is upwards

$u = 21$

$a = -9.8$

$s = ut + \frac{1}{2}at^2$

$s_2 = 21t - 4.9t^2$

The height of the tower is 63 m.

As B is moving upwards, the acceleration due to gravity is negative.

You now have expressions for s_1 and s_2 in terms of t.

$s_1 + s_2 = 63$

$4.9t^2 + (21t - 4.9t^2) = 63$

$21t = 63$

$t = 3$

$s_2 = 21t - 4.9t^2$

$= 21 \times 3 - 4.9 \times 3^2 = 18.9$

The balls collide 19 m from the bottom of the tower, to two significant figures.

Adding together the two distances gives the height of the tower. You can write this as an equation in t.

You have found t but you were asked for the distance from the bottom of the tower. Substitute your value for t into your equation for s_2.

Exercise 9E

1 A cliff diver jumps from a point 28 m above the surface of the water. Modelling the diver as a particle moving freely under gravity with initial velocity 0, find:
 a the time taken for the diver to hit the water
 b the speed of the diver when he hits the water.

2 A particle is projected vertically upwards with speed $20\,\text{m s}^{-1}$ from a point on the ground. Find the time of flight of the particle.

3 A ball is thrown vertically downward from the top of a tower with speed $18\,\text{m s}^{-1}$. It reaches the ground in 1.6 s. Find the height of the tower.

4 A pebble is catapulted vertically upwards with speed $24\,\text{m s}^{-1}$. Find:
 a the greatest height above the point of projection reached by the pebble
 b the time taken to reach this height.

5 A ball is projected upwards from a point which is 4 m above the ground with speed $18\,\text{m s}^{-1}$. Find:
 a the speed of the ball when it is 15 m above its point of projection
 b the speed with which the ball hits the ground.

6 A particle P is projected vertically downwards from a point 80 m above the ground with speed $4\,\text{m s}^{-1}$. Find:
 a the speed with which P hits the ground
 b the time P takes to reach the ground.

7 A particle P is projected vertically upwards from a point X. Five seconds later P is moving downwards with speed $10\,\text{m s}^{-1}$. Find:
 a the speed of projection of P
 b the greatest height above X attained by P during its motion.

8 A ball is thrown vertically upwards with speed $21\,\text{m s}^{-1}$. It hits the ground 4.5 s later. Find the height above the ground from which the ball was thrown.

9 A stone is thrown vertically upward from a point which is 3 m above the ground, with speed $16\,\text{m s}^{-1}$. Find:
 a the time of flight of the stone
 b the total distance travelled by the stone.

(P) 10 A particle is projected vertically upwards with speed $24.5\,\text{m s}^{-1}$. Find the total time for which it is 21 m or more above its point of projection.

(E/P) 11 A particle is projected vertically upwards from a point O with speed $u\,\text{m s}^{-1}$. Two seconds later it is still moving upwards and its speed is $\frac{1}{3}u\,\text{m s}^{-1}$. Find:

Problem-solving

Use $v = u + at$ and substitute $v = \frac{1}{3}u$.

 a the value of u **(3 marks)**

 b the time from the instant that the particle leaves O to the instant that it returns to O. **(4 marks)**

(E/P) **12** A ball A is thrown vertically downwards with speed $5\,\mathrm{m\,s^{-1}}$ from the top of a tower block $46\,\mathrm{m}$ above the ground. At the same time as A is thrown downwards, another ball B is thrown vertically upwards from the ground with speed $18\,\mathrm{m\,s^{-1}}$. The balls collide. Find the distance of the point where A and B collide from the point where A was thrown. **(5 marks)**

(E/P) **13** A ball is released from rest at a point which is $10\,\mathrm{m}$ above a wooden floor. Each time the ball strikes the floor, it rebounds with three-quarters of the speed with which it strikes the floor. Find the greatest height above the floor reached by the ball

> **Problem-solving**
>
> Consider each bounce as a separate motion.

　　a the first time it rebounds from the floor **(3 marks)**

　　b the second time it rebounds from the floor. **(4 marks)**

Challenge

1 A particle P is projected vertically upwards from a point O with speed $12\,\mathrm{m\,s^{-1}}$. One second after P has been projected from O, another particle Q is projected vertically upwards from O with speed $20\,\mathrm{m\,s^{-1}}$. Find: **a** the time between the instant that P is projected from O and the instant when P and Q collide, **b** the distance of the point where P and Q collide from O.

2 A stone is dropped from the top of a building and two seconds later another stone is thrown vertically downwards at a speed of $25\,\mathrm{m\,s^{-1}}$. Both stones reach the ground at the same time. Find the height of the building.

Mixed exercise (9)

1 A car accelerates in a straight line at a constant rate, starting from rest at a point A and reaching a velocity of $45\,\mathrm{km\,h^{-1}}$ in $20\,\mathrm{s}$. This velocity is then maintained and the car passes a point B 3 minutes after leaving A.

　　a Sketch a velocity–time graph to illustrate the motion of the car.

　　b Find the displacement of the car from its starting point after 3 minutes.

2 A particle is moving on an axis Ox. From time $t = 0\,\mathrm{s}$ to time $t = 32\,\mathrm{s}$, the particle is travelling with constant velocity $15\,\mathrm{m\,s^{-1}}$. The particle then decelerates from $15\,\mathrm{m\,s^{-1}}$ to rest in T seconds.

　　a Sketch a velocity–time graph to illustrate the motion of the particle.

　　The total distance travelled by the particle is $570\,\mathrm{m}$.

　　b Find the value of T.

　　c Sketch a displacement–time graph illustrating the motion of the particle.

(P) **3** The velocity–time graph represents the motion of a particle moving in a straight line accelerating from velocity u at time 0 to velocity v at time t.

　　a Use the graph to show that:

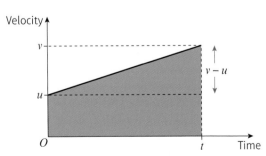

　　　i $v = u + at$　　**ii** $s = \left(\dfrac{u + v}{2}\right)t$

　　b Hence show that:

　　　i $v^2 = u^2 + 2as$　　**ii** $s = ut + \tfrac{1}{2}at^2$　　**iii** $s = vt - \tfrac{1}{2}at^2$

(P) 4 The diagram is a velocity–time graph representing the motion of a cyclist along a straight road. At time $t = 0$ s, the cyclist is moving with velocity u m s^{-1}. The velocity is maintained until time $t = 15$ s, when she slows down with constant deceleration, coming to rest when $t = 23$ s. The total distance she travels in 23 s is 152 m. Find the value of u.

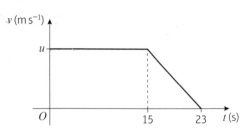

5 A car travelling on a straight road slows down with constant deceleration. The car passes a road sign with velocity 40 km h^{-1} and a post box with velocity of 24 km h^{-1}. The distance between the road sign and the post box is 240 m. Find, in m s^{-2}, the deceleration of the car.

6 A particle P is moving along the x-axis with constant deceleration 2.5 m s^{-2}. At time $t = 0$ s, P passes through the origin with velocity 20 m s^{-1} in the direction of x increasing. At time $t = 12$ s, P is at the point A. Find:

a the distance OA b the total distance P travels in 12 s.

7 A ball is thrown vertically downward from the top of a tower with speed 6 m s^{-1}. The ball strikes the ground with speed 25 m s^{-1}. Find the time the ball takes to move from the top of the tower to the ground.

8 A child drops a ball from a point at the top of a cliff which is 82 m above the sea. The ball is initially at rest. Find:

a the time taken for the ball to reach the sea b the speed with which the ball hits the sea.

c State one physical factor which has been ignored in making your calculation.

(P) 9 A particle moves 451 m in a straight line. The diagram shows a speed–time graph illustrating the motion of the particle. The particle starts at rest and accelerates at a constant rate for 8 s reaching a speed of $2u$ m s^{-1}. The particle then travels at a constant speed for 12 seconds before decelerating uniformly, reaching a speed of u m s^{-1} at time $t = 26$ s. Find:

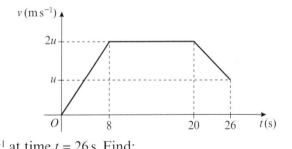

a the value of u

b the distance moved by the particle while its speed is less than u m s^{-1}.

(E/P) 10 A train is travelling with constant acceleration along a straight track. At time $t = 0$ s, the train passes a point O travelling with velocity 18 m s^{-1}. At time $t = 12$ s, the train passes a point P travelling with velocity 24 m s^{-1}. At time $t = 20$ s, the train passes a point Q. Find:

a the speed of the train at Q **(5 marks)**

b the distance from P to Q. **(2 marks)**

(E) 11 A particle moves along a straight line, from a point X to a point Y, with constant acceleration. The distance from X to Y is 104 m. The particle takes 8 s to move from X to Y and the speed of the particle at Y is 18 m s^{-1}. Find:

a the speed of the particle at X (**3 marks**)

b the acceleration of the particle. (**2 marks**)

The particle continues to move with the same acceleration until it reaches a point Z. At Z the speed of the particle is three times the speed of the particle at X.

c Find the distance XZ. (**4 marks**)

(E) **12** A pebble is projected vertically upwards with speed $21\,\text{m s}^{-1}$ from a point $32\,\text{m}$ above the ground. Find:

a the speed with which the pebble strikes the ground (**3 marks**)

b the total time for which the pebble is more than $40\,\text{m}$ above the ground. (**4 marks**)

c Sketch a velocity–time graph for the motion of the pebble from the instant it is projected to the instant it hits the ground, showing the values of t at any points where the graph intercepts the horizontal axis. (**4 marks**)

(E) **13** A car is moving along a straight road with uniform acceleration. The car passes a checkpoint A with speed $12\,\text{m s}^{-1}$ and another checkpoint C with speed $32\,\text{m s}^{-1}$. The distance between A and C is $1100\,\text{m}$.

a Find the time taken by the car to move from A to C. (**2 marks**)

b Given that B is the midpoint of AC, find the speed with which the car passes B. (**2 marks**)

(E/P) **14** A particle is projected vertically upwards with a speed of $30\,\text{m s}^{-1}$ from a point A. The point B is h metres above A. The particle moves freely under gravity and is above B for a time $2.4\,\text{s}$. Calculate the value of h. (**5 marks**)

(E/P) **15** Two cars A and B are moving in the same direction along a straight horizontal road. At time $t = 0$, they are side by side, passing a point O on the road. Car A travels at a constant speed of $30\,\text{m s}^{-1}$. Car B passes O with a speed of $20\,\text{m s}^{-1}$, and has constant acceleration of $4\,\text{m s}^{-2}$. Find:

a the speed of B when it has travelled $78\,\text{m}$ from O (**2 marks**)

b the distance from O of A when B is $78\,\text{m}$ from O (**3 marks**)

c the time when B overtakes A. (**4 marks**)

(E/P) **16** A car is being driven on a straight stretch of motorway at a constant velocity of $34\,\text{m s}^{-1}$, when it passes a velocity restriction sign S warning of road works ahead and requiring speeds to be reduced to $22\,\text{m s}^{-1}$. The driver continues at her velocity for $2\,\text{s}$ after passing S. She then reduces her velocity to $22\,\text{m s}^{-1}$ with constant deceleration of $3\,\text{m s}^{-2}$, and continues at the lower velocity.

a Draw a velocity–time graph to illustrate the motion of the car after it passes S. (**2 marks**)

b Find the shortest distance before the road works that S should be placed on the road to ensure that a car driven in this way has had its velocity reduced to $22\,\text{m s}^{-1}$ by the time it reaches the start of the road works. (**4 marks**)

(E/P) **17** A train starts from rest at station A and accelerates uniformly at $3x\,\text{m s}^{-2}$ until it reaches a velocity of $30\,\text{m s}^{-1}$. For the next T seconds the train maintains this constant velocity. The train then decelerates uniformly at $x\,\text{m s}^{-2}$ until it comes to rest at a station B. The distance between the stations is $6\,\text{km}$ and the time taken from A to B is 5 minutes.

a Sketch a velocity–time graph to illustrate this journey. **(2 marks)**

b Show that $\dfrac{40}{x} + T = 300$. **(4 marks)**

c Find the value of T and the value of x. **(2 marks)**

d Calculate the distance the train travels at constant velocity. **(2 marks)**

e Calculate the time taken from leaving A until reaching the point halfway between the stations. **(3 marks)**

Challenge

A ball is projected vertically upwards with speed $10 \, \text{m s}^{-1}$ from a point X, which is 50 m above the ground. T seconds after the first ball is projected upwards, a second ball is dropped from X. Initially the second ball is at rest. The balls collide 25 m above the ground. Find the value of T.

Summary of key points

1 Velocity is the **rate of change** of displacement.

On a displacement–time graph the **gradient** represents the velocity.

If the displacement–time graph is a straight line, then the velocity is constant.

2 Average velocity $= \dfrac{\text{displacement from starting point}}{\text{time taken}}$

3 Average speed $= \dfrac{\text{total distance travelled}}{\text{time taken}}$

4 Acceleration is the **rate of change** of velocity.

In a velocity–time graph the **gradient** represents the acceleration.

If the velocity–time graph is a straight line, then the acceleration is constant.

5 The area between a velocity–time graph and the horizontal axis represents the distance travelled.

For motion in a straight line with positive velocity, the area under the velocity–time graph up to a point t represents the displacement at time t.

6 You need to be able to use and to derive the five formulae for solving problems about particles moving in a straight line with constant acceleration.

- $v = u + at$
- $s = \left(\dfrac{u + v}{2}\right)t$
- $v^2 = u^2 + 2as$
- $s = ut + \frac{1}{2}at^2$
- $s = vt - \frac{1}{2}at^2$

7 The force of **gravity** causes all objects to accelerate towards the earth. If you ignore the effects of air resistance, this acceleration is constant. It does not depend on the mass of the object.

8 An object moving vertically in a straight line can be modelled as a particle with a constant downward acceleration of $g = 9.8 \, \text{m s}^{-2}$.

Objectives

After completing this chapter you should be able to:

● Draw force diagrams and calculate resultant forces   → pages 157–159
● Understand and use Newton's first law   → pages 157–159
● Calculate resultant forces by adding vectors   → pages 160–162
● Understand and use Newton's second law, $F = ma$ → pages 162–166
● Apply Newton's second law to vector forces and acceleration

                                                                        → pages 166–169

● Understand and use Newton's third law   → pages 169–172
● Solve problems involving connected particles   → pages 169–177

Prior knowledge check

1   Calculate:
    a $(2\mathbf{i} + \mathbf{j}) + (3\mathbf{i} - 4\mathbf{j})$ **b** $(-\mathbf{i} + 3\mathbf{j}) - (3\mathbf{i} - \mathbf{j})$
                                                             ← Section 8.4

2   The diagram shows a
    right-angled triangle.
    Work out:
    a the length of the
        hypotenuse
    b the size of the angle a.
    Give your answers correct to 1 d.p.   ← GCSE Mathematics

3   A car starts from rest and accelerates constantly at
    $1.5\ \text{m s}^{-2}$.
    a Work out the velocity of the car after 12 seconds.

    After 12 seconds, the driver brakes, causing the car to
    decelerate at a constant rate of $1\ \text{m s}^{-2}$.
    b Calculate the distance the car travels from the
        instant the driver brakes until the car comes to rest.
                                                        ← Chapter 9

The weight of an
air–sea rescue crew
man is balanced by
the tension in the
cable. By modelling
the forces in this
situation, you can
calculate how strong
the cable needs to be.
                    → Exercise 10A Q5

10.1 Force diagrams

A force diagram is a diagram showing all the forces acting on an object. Each force is shown as an arrow pointing in the direction in which the force acts. Force diagrams are used to model problems involving forces.

Example 1

A block of weight W is being pulled to the right by a force, P, across a rough horizontal plane. Draw a force diagram to show all the forces acting on the block.

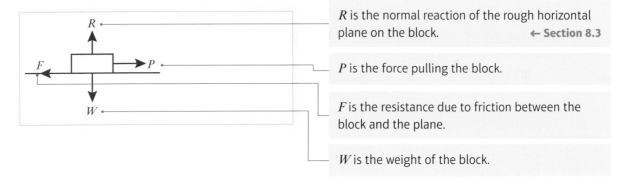

R is the normal reaction of the rough horizontal plane on the block. ← Section 8.3

P is the force pulling the block.

F is the resistance due to friction between the block and the plane.

W is the weight of the block.

When the forces acting upon an object are balanced, the object is said to be in equilibrium.

■ **Newton's first law of motion states that an object at rest will stay at rest and that an object moving with constant velocity will continue to move with constant velocity unless an unbalanced force acts on the object.**

Watch out Constant velocity means that neither the speed nor the direction is changing.

When there is more than one force acting on an object you can resolve the forces in a certain direction to find the resultant force in that direction. The direction you are resolving in becomes the positive direction. You add forces acting in this direction and subtract forces acting in the opposite direction.

In your answers, you can use the letter R, together with an arrow, R(↑), to indicate the direction in which you are resolving the forces.

In this section you will only resolve forces that are horizontal or vertical.

■ **A resultant force acting on an object will cause the object to accelerate in the same direction as the resultant force.**

Example 2

The diagram shows the forces acting on a particle.
a Draw a force diagram to represent the resultant force.
b Describe the motion of the particle.

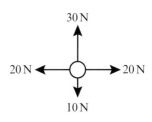

a 20 N

b The particle is accelerating upwards.

R(→): 20 − 20 = 0 so the horizontal forces are balanced.

R(↑): 30 − 10 = 20 so the resultant force is 20 N upwards.

Exercise 10A

1 A box is at rest on a horizontal table. Draw a force diagram to show all the forces acting on the box.

2 A trapeze bar is suspended motionless from the ceiling by two ropes. Draw a force diagram to show the forces acting on the ropes and the trapeze bar.

3 Ignoring air resistance, draw a diagram to show the forces acting on an apple as it falls from a tree.

4 A car's engine applies a force parallel to the surface of a horizontal road that causes the car to move with constant velocity. Considering the resistance to motion, draw a diagram to show the forces acting on the car.

5 An air–sea rescue crew member is suspended motionless from a helicopter. Ignoring air resistance, show all the forces acting on him.

(P) 6 A satellite orbits the Earth at constant speed. State, with a reason, whether any resultant force is acting on the satellite.

Problem-solving

Consider the velocity of the satellite as it orbits the Earth.

7 A particle of weight 5 N sits at rest on a horizontal plane. State the value of the normal reaction acting on the particle.

8 Given that each of the particles is stationary, work out the value of P:

a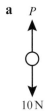

b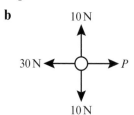

c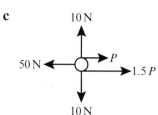

9 A hoist lifts a platform vertically at constant velocity as shown in the diagram.

a Ignoring air resistance, work out the tension, T in each rope.

The tension in each rope is reduced by 50 N.

b Describe the resulting motion of the platform.

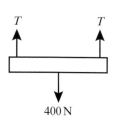

10 The diagram shows a particle acted on by a set of forces.
Given that the particle is at rest, find the value of p and the value
of q.

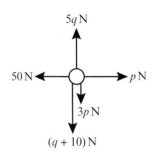

(P) 11 Given that the particle in this diagram is moving with constant
velocity, v, find the values of P and Q.

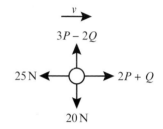

Problem-solving

Set up two simultaneous
equations.

12 Each diagram shows the forces acting on a particle.
  i Work out the size and direction of the resultant force.
  ii Describe the motion of the particle.

a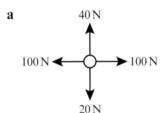

b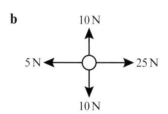

13 A truck is moving along a horizontal level road. The truck's engine provides a forward thrust of
10 000 N. The total resistance is modelled as a constant force of magnitude 1600 N.

  a Modelling the truck as a particle, draw a force diagram to show the forces acting on the
  truck.

  b Calculate the resultant force acting on the truck.

(P) 14 A car is moving along a horizontal level road. The car's engine provides a constant driving
force. The motion of the car is opposed by a constant resistance.

  a Modelling the car as a particle, draw a force diagram to show the forces acting on the car.

  b Given that the resultant force acting on the car is 4200 N in the direction of motion, and
  that the magnitude of the driving force is eight times the magnitude of the resistance force,
  calculate the magnitude of the resistance.

Problem-solving

Use algebra to describe the relationship
between the driving force and the
resistance.

10.2 Forces as vectors

You can write forces as vectors using **i–j** notation or as column vectors.

■ **You can find the resultant of two or more forces given as vectors by adding the vectors.**

> **Links** If two forces $(p\mathbf{i} + q\mathbf{j})$ N and $(r\mathbf{i} + s\mathbf{j})$ N are acting on a particle, the resultant force will be $((p + r)\mathbf{i} + (q + s)\mathbf{j})$ N. ← **Pure Year 1, Section 11.2**

When a particle is in equilibrium the resultant vector force will be $0\mathbf{i} + 0\mathbf{j}$.

Example 3

The forces $2\mathbf{i} + 3\mathbf{j}$, $4\mathbf{i} - \mathbf{j}$, $-3\mathbf{i} + 2\mathbf{j}$ and $a\mathbf{i} + b\mathbf{j}$ act on an object which is in equilibrium. Find the values of a and b.

$(2\mathbf{i} + 3\mathbf{j}) + (4\mathbf{i} - \mathbf{j}) + (-3\mathbf{i} + 2\mathbf{j}) + (a\mathbf{i} + b\mathbf{j}) = 0$

$(2 + 4 - 3 + a)\mathbf{i} + (3 - 1 + 2 + b)\mathbf{j} = 0$

$\Rightarrow \quad 3 + a = 0 \quad \text{and} \quad 4 + b = 0$

$\Rightarrow \quad\quad a = -3 \quad \text{and} \quad b = -4$

> If an object is in equilibrium then the resultant force will be zero.

> You can consider the **i** and **j** components separately because they are perpendicular.

Example 4

In this question **i** represents the unit vector due east, and **j** represents the unit vector due north. A particle begins at rest at the origin. It is acted on by three forces $(2\mathbf{i} + \mathbf{j})$ N, $(3\mathbf{i} - 2\mathbf{j})$ N and $(-\mathbf{i} + 4\mathbf{j})$ N.

a Find the resultant force in the form $p\mathbf{i} + q\mathbf{j}$.

b Work out the magnitude and bearing of the resultant force.

c Describe the motion of the particle.

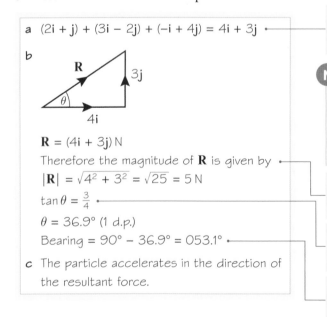

a $(2\mathbf{i} + \mathbf{j}) + (3\mathbf{i} - 2\mathbf{j}) + (-\mathbf{i} + 4\mathbf{j}) = 4\mathbf{i} + 3\mathbf{j}$

b

$\mathbf{R} = (4\mathbf{i} + 3\mathbf{j})$ N

Therefore the magnitude of **R** is given by

$|\mathbf{R}| = \sqrt{4^2 + 3^2} = \sqrt{25} = 5$ N

$\tan\theta = \frac{3}{4}$

$\theta = 36.9°$ (1 d.p.)

Bearing $= 90° - 36.9° = 053.1°$

c The particle accelerates in the direction of the resultant force.

> Add together the **i**-components and the **j**-components.

> **Notation** The unit vector **i** is usually taken to be due east or the positive x-direction. The unit vector **j** is usually taken to be due north or the positive y-direction. Questions involving finding bearings will often specify this.

> Use Pythagoras' theorem to find the magnitude of the resultant.

> Use $\tan\theta = \dfrac{\text{opp}}{\text{adj}}$

> Bearings are measured clockwise from north so subtract θ from 90°.

Exercise 10B

1 In each part of the question a particle is acted upon by the forces given. Work out the resultant force acting on the particle.

    a $(-\mathbf{i} + 3\mathbf{j})$ N and $(4\mathbf{i} - \mathbf{j})$ N   **b** $\begin{pmatrix} 5 \\ 3 \end{pmatrix}$ N and $\begin{pmatrix} -3 \\ -6 \end{pmatrix}$ N

Notation $\begin{pmatrix} 5 \\ 3 \end{pmatrix}$ N is the same as $(5\mathbf{i} + 3\mathbf{j})$ N.

    c $(\mathbf{i} + \mathbf{j})$ N, $(5\mathbf{i} - 3\mathbf{j})$ N and $(-2\mathbf{i} - \mathbf{j})$ N   **d** $\begin{pmatrix} -1 \\ 4 \end{pmatrix}$ N, $\begin{pmatrix} 6 \\ 0 \end{pmatrix}$ N and $\begin{pmatrix} -2 \\ -7 \end{pmatrix}$ N

2 An object is in equilibrium at O under the action of three forces $\mathbf{F}_1$, $\mathbf{F}_2$ and $\mathbf{F}_3$. Find $\mathbf{F}_3$ in these cases.

    a $\mathbf{F}_1 = (2\mathbf{i} + 7\mathbf{j})$ and $\mathbf{F}_2 = (-3\mathbf{i} + \mathbf{j})$ **b** $\mathbf{F}_1 = (3\mathbf{i} - 4\mathbf{j})$ and $\mathbf{F}_2 = (2\mathbf{i} + 3\mathbf{j})$

(P) 3 The forces $\begin{pmatrix} a \\ 2b \end{pmatrix}$ N, $\begin{pmatrix} -2a \\ -b \end{pmatrix}$ N and $\begin{pmatrix} 3 \\ -4 \end{pmatrix}$ N act on an object which is in equilibrium.

    Find the values of a and b.

4 For each force find:

    i the magnitude of the force   **ii** the angle the force makes with $\mathbf{i}$

    a $(3\mathbf{i} + 4\mathbf{j})$ N   **b** $(5\mathbf{i} - \mathbf{j})$ N   **c** $(-2\mathbf{i} + 3\mathbf{j})$ N   **d** $\begin{pmatrix} -1 \\ -1 \end{pmatrix}$ N

5 In this question, $\mathbf{i}$ represents the unit vector due east, and $\mathbf{j}$ represents the unit vector due north. A particle is acted upon by forces of:

    a $(-2\mathbf{i} + \mathbf{j})$ N, $(5\mathbf{i} + 2\mathbf{j})$ N and $(-\mathbf{i} - 4\mathbf{j})$ N   **b** $(-2\mathbf{i} + \mathbf{j})$ N, $(2\mathbf{i} - 3\mathbf{j})$ N and $(3\mathbf{i} + 6\mathbf{j})$ N

    Work out:

    i the resultant vector

    ii the magnitude of the resultant vector

    iii the bearing of the resultant vector.

(P) 6 The forces $(a\mathbf{i} - b\mathbf{j})$ N, $(b\mathbf{i} + a\mathbf{j})$ N and $(-4\mathbf{i} - 2\mathbf{j})$ N act on an object which is in equilibrium. Find the values of a and b.

Problem-solving

Use the $\mathbf{i}$ components and the $\mathbf{j}$ components to set up and solve two simultaneous equations.

(P) 7 The forces $(2a\mathbf{i} + 2b\mathbf{j})$ N, $(-5b\mathbf{i} + 3a\mathbf{j})$ N and $(-11\mathbf{i} - 7\mathbf{j})$ N act on an object which is in equilibrium. Find the values of a and b.

8 Three forces $\mathbf{F}_1$, $\mathbf{F}_2$ and $\mathbf{F}_3$ act on a particle. $\mathbf{F}_1 = (-3\mathbf{i} + 7\mathbf{j})$ N, $\mathbf{F}_2 = (\mathbf{i} - \mathbf{j})$ N and $\mathbf{F}_3 = (p\mathbf{i} + q\mathbf{j})$ N.

    a Given that this particle is in equilibrium, determine the value of p and the value of q.

    The resultant of the forces $\mathbf{F}_1$ and $\mathbf{F}_2$ is $\mathbf{R}$.

    b Calculate, in N, the magnitude of $\mathbf{R}$.

    c Calculate, to the nearest degree, the angle between the line of action of $\mathbf{R}$ and the vector $\mathbf{j}$.

E/P 9 A particle is acted upon by two forces $\mathbf{F}_1$ and $\mathbf{F}_2$, given by $\mathbf{F}_1 = (3\mathbf{i} - 2\mathbf{j})$ N and $\mathbf{F}_2 = (a\mathbf{i} + 2a\mathbf{j})$ N, where a is a positive constant.

    a Find the angle between $\mathbf{F}_2$ and $\mathbf{i}$.   **(2 marks)**

    The resultant of $\mathbf{F}_1$ and $\mathbf{F}_2$ is $\mathbf{R}$.

    b Given that $\mathbf{R}$ is parallel to $13\mathbf{i} + 10\mathbf{j}$, find the value of a.   **(4 marks)**

E 10 Three forces $\mathbf{F}_1$, $\mathbf{F}_2$ and $\mathbf{F}_3$ acting on a particle P are given by the vectors $\mathbf{F}_1 = \begin{pmatrix} -7 \\ -4 \end{pmatrix}$ N,

    $\mathbf{F}_2 = \begin{pmatrix} 4 \\ 2 \end{pmatrix}$ N and $\mathbf{F}_3 = \begin{pmatrix} a \\ b \end{pmatrix}$ N, where a and b are constants.

    Given that P is in equilibrium,

    a find the value of a and the value of b.   **(3 marks)**

    b The force $\mathbf{F}_1$ is now removed. The resultant of $\mathbf{F}_2$ and $\mathbf{F}_3$ is $\mathbf{R}$. Find:

      i the magnitude of $\mathbf{R}$ **(2 marks)**

      ii the angle, to the nearest degree, that the direction of $\mathbf{R}$ makes with the horizontal.

                                                                             (3 marks)

Challenge

An object is acted upon by a horizontal force of $10\mathbf{i}$ N and a vertical force $a\mathbf{j}$ N as shown in the diagram. The resultant of the two forces acts in the direction $60°$ to the horizontal. Work out the value of a and the magnitude of the resultant force.

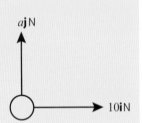

10.3 Forces and acceleration

A non-zero resultant force that acts on a particle will cause the particle to accelerate in the direction of the resultant force.

- **Newton's second law of motion states that the force needed to accelerate a particle is equal to the product of the mass of the particle and the acceleration produced: $F = ma$.**

A force of 1 N will accelerate a mass of 1 kg at a rate of 1 m s^{-2}. If a force F N acts on a particle of mass m kg causing it to accelerate at a m s^{-2}, the **equation of motion** for the particle is $F = ma$.

Gravity is the force between any object and the Earth. The force due to gravity acting on an object is called the **weight** of the object, and it acts vertically downwards. A body falling freely experiences an acceleration of $g = 9.8$ m s^{-2}. Using the relationship $F = ma$ you can write the equation of motion for a body of mass m kg with weight W N.

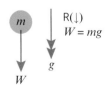

- **$W = mg$**

Example 5

Find the acceleration when a particle of mass 1.5 kg is acted on by a resultant force of 6 N.

$F = ma$

$6 = 1.5a$ ——————————————— Substitute the values you know and solve the equation to find a.

$a = 4$

The acceleration is $4\,\mathrm{m\,s^{-2}}$.

Example 6

In each of these diagrams the body is accelerating as shown. Find the magnitudes of the unknown forces X and Y.

a

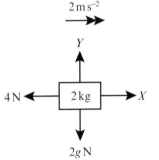

b

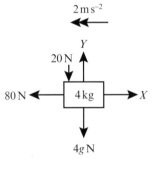

a R($\rightarrow$), $X - 4 = 2 \times 2$ ————— R($\rightarrow$), means that you are finding the resultant force in the horizontal direction, in the direction of the arrow. The arrow is in the positive direction.

$\qquad X = 8\,\mathrm{N}$

R($\uparrow$), $Y - 2g = 2 \times 0$

$\qquad Y = 2g$

$\qquad\quad = 19.6\,\mathrm{N}$

This resultant force causes an acceleration of $2\,\mathrm{m\,s^{-2}}$. Use $F = ma$.

b R($\leftarrow$), $80 - X = 4 \times 2$

$\qquad X = 72\,\mathrm{N}$

It is usually easier to take the positive direction as the direction of the acceleration.

R($\uparrow$), $Y - 20 - 4g = 4 \times 0$ ————— There is no vertical acceleration, so $a = 0$.

$\qquad Y = 20 + (4 \times 9.8)$

$\qquad\quad = 59.2\,\mathrm{N}$ (3 s.f.)

Example 7

A body of mass 5 kg is pulled along a rough horizontal table by a horizontal force of magnitude 20 N against a constant friction force of magnitude 4 N. Given that the body is initially at rest, find:

a the acceleration of the body

b the distance travelled by the body in the first 4 seconds

c the magnitude of the normal reaction between the body and the table.

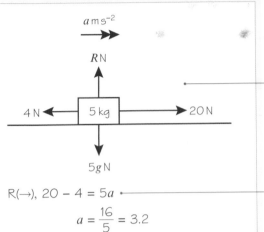

a

Draw a diagram showing all the forces and the acceleration.

$R(\rightarrow)$, $20 - 4 = 5a$

Resolve horizontally, taking the positive direction as the direction of the acceleration, and write down an equation of motion for the body.

$a = \dfrac{16}{5} = 3.2$

The body accelerates at $3.2\,\mathrm{m\,s^{-2}}$.

b $s = ut + \dfrac{1}{2}at^2$

Since the acceleration is constant.

$s = (0 \times 4) + \dfrac{1}{2} \times 3.2 \times 4^2$

Substitute in the values.

$= 25.6$

The body moves a distance of $25.6\,\mathrm{m}$.

c $R(\uparrow)$, $R - 5g = 5 \times 0 = 0$

Resolve vertically. Since the body is moving horizontally $a = 0$, so the right-hand side of the equation of motion is 0.

$R = 5g = 5 \times 9.8 = 49\,\mathrm{N}$

The normal reaction has magnitude $49\,\mathrm{N}$.

Exercise 10C

1 Find the acceleration when a particle of mass $400\,\mathrm{kg}$ is acted on by a resultant force of $120\,\mathrm{N}$.

2 Find the weight in newtons of a particle of mass $4\,\mathrm{kg}$.

3 An object moving on a rough surface experiences a constant frictional force of $30\,\mathrm{N}$ which decelerates it at a rate of $1.2\,\mathrm{m\,s^{-2}}$. Find the mass of the object.

(P) 4 An astronaut weighs $735\,\mathrm{N}$ on the earth and $120\,\mathrm{N}$ on the moon. Work out the value of acceleration due to gravity on the moon.

Problem-solving

Start by finding the mass of the astronaut.

5 In each scenario, the forces acting on the body cause it to accelerate as shown. Find the magnitude of the unknown force.

a

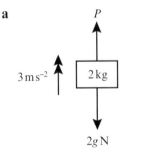

b

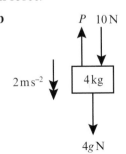

6 In each situation, the forces acting on the body cause it to accelerate as shown. In each case find the mass of the body, m.

a

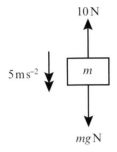

b

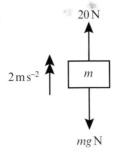

7 In each situation, the forces acting on the body cause it to accelerate as shown with magnitude $a\,\mathrm{m\,s^{-2}}$. In each case find the value of a.

a

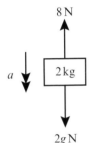

b

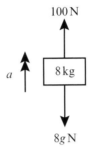

(P) 8 A force of 10 N acts upon a particle of mass 3 kg causing it to accelerate at $2\,\mathrm{m\,s^{-2}}$ along a rough horizontal plane. Calculate the value of the force due to friction.

Problem-solving

Draw a force diagram showing all the forces acting on the particle.

(E/P) 9 A lift of mass 500 kg is lowered or raised by means of a metal cable attached to its top. The lift contains passengers whose total mass is 300 kg. The lift starts from rest and accelerates at a constant rate, reaching a speed of $3\,\mathrm{m\,s^{-1}}$ after moving a distance of 5 m. Find:

a the acceleration of the lift **(3 marks)**

Hint Use $v^2 = u^2 + 2as$.

b the tension in the cable if the lift is moving vertically downwards **(2 marks)**

c the tension in the cable if the lift is moving vertically upwards. **(2 marks)**

(E) 10 A trolley of mass 50 kg is pulled from rest in a straight line along a horizontal path by means of a horizontal rope attached to its front end. The trolley accelerates at a constant rate and after 2 s its speed is $1\,\mathrm{m\,s^{-1}}$. As it moves, the trolley experiences a resistance to motion of magnitude 20 N. Find:

a the acceleration of the trolley **(3 marks)**

b the tension in the rope. **(2 marks)**

(E/P) 11 The engine of a van of mass 400 kg cuts out when it is moving along a straight horizontal road with speed $16\,\mathrm{m\,s^{-1}}$. The van comes to rest without the brakes being applied.

In a model of the situation it is assumed that the van is subject to a resistive force which has constant magnitude of 200 N.

a Find how long it takes the van to stop. **(3 marks)**

b Find how far the van travels before it stops. **(2 marks)**

c Comment on the suitability of the modelling assumption. **(1 mark)**

A small stone of mass 400 g is projected vertically upwards from the bottom of a pond full of water with speed 10 m s⁻¹. As the stone moves through the water it experiences a constant resistance of magnitude 3 N. Assuming that the stone does not reach the surface of the pond, find:

a the greatest height above the bottom of the pond that the stone reaches

b the speed of the stone as it hits the bottom of the pond on its return

c the total time taken for the stone to return to its initial position on the bottom of the pond.

10.4 Motion in 2 dimensions

■ **You can use F = *m*a to solve problems involving vector forces acting on particles.**

Notation In this version of the equation of motion, **F** and **a** are vectors. You can write acceleration as a 2D vector in the form $(p\mathbf{i} + q\mathbf{j})$ m s⁻² or $\begin{pmatrix} p \\ q \end{pmatrix}$ m s⁻².

Example 8

In this question **i** represents the unit vector due east, and **j** represents the unit vector due north.

A resultant force of $(3\mathbf{i} + 8\mathbf{j})$ N acts upon a particle of mass 0.5 kg.

a Find the acceleration of the particle in the form $(p\mathbf{i} + q\mathbf{j})$ m s⁻².

b Find the magnitude and bearing of the acceleration of the particle.

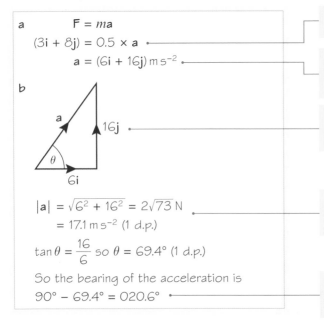

a $\mathbf{F} = m\mathbf{a}$

$(3\mathbf{i} + 8\mathbf{j}) = 0.5 \times \mathbf{a}$ ·

$\mathbf{a} = (6\mathbf{i} + 16\mathbf{j})$ m s⁻²

Write the vector equation of motion.

To divide $(3\mathbf{i} + 8\mathbf{j})$ by 0.5, you divide each component by 0.5.

b

Draw a diagram to represent the acceleration vector.

$|\mathbf{a}| = \sqrt{6^2 + 16^2} = 2\sqrt{73}$ N

$= 17.1$ m s⁻² (1 d.p.)

Use Pythagoras' theorem to work out the magnitude of the acceleration vector.

$\tan\theta = \dfrac{16}{6}$ so $\theta = 69.4°$ (1 d.p.)

So the bearing of the acceleration is

$90° - 69.4° = 020.6°$ ·

Remember bearings are always measured clockwise from north.

Example 9

Forces $\mathbf{F}_1 = (2\mathbf{i} + 4\mathbf{j})$ N, $\mathbf{F}_2 = (-5\mathbf{i} + 4\mathbf{j})$ N, and $\mathbf{F}_3 = (6\mathbf{i} - 5\mathbf{j})$ N act on a particle of mass 3 kg. Find the acceleration of the particle.

Resultant force
$$= \mathbf{F}_1 + \mathbf{F}_2 + \mathbf{F}_3$$ ——— Add the vectors to find the resultant force.
$$= (2\mathbf{i} + 4\mathbf{j}) + (-5\mathbf{i} + 4\mathbf{j}) + (6\mathbf{i} - 5\mathbf{j})$$
$$= 3\mathbf{i} + 3\mathbf{j}$$
$$3\mathbf{i} + 3\mathbf{j} = 3\mathbf{a} \implies \mathbf{a} = (\mathbf{i} + \mathbf{j}) \, \text{m s}^{-2}$$ ——— Use $\mathbf{F} = m\mathbf{a}$.

Example 10

A boat is modelled as a particle of mass 60 kg being acted on by three forces:

$$\mathbf{F}_1 = \begin{pmatrix} 80 \\ 50 \end{pmatrix} \text{N} \qquad \mathbf{F}_2 = \begin{pmatrix} 10p \\ 20q \end{pmatrix} \text{N} \qquad \mathbf{F}_3 = \begin{pmatrix} -75 \\ 100 \end{pmatrix} \text{N}$$

Given that the boat is accelerating at a rate of $\begin{pmatrix} 0.8 \\ -1.5 \end{pmatrix}$ m s^{-2}, find the values of p and q.

Resultant force
$$= \mathbf{F}_1 + \mathbf{F}_2 + \mathbf{F}_3$$

$$= \begin{pmatrix} 80 \\ 50 \end{pmatrix} + \begin{pmatrix} 10p \\ 20q \end{pmatrix} + \begin{pmatrix} -75 \\ 100 \end{pmatrix}$$

$$= \begin{pmatrix} 5 + 10p \\ 150 + 20q \end{pmatrix} \text{N}$$ ——— Find the resultant force acting on the boat in terms of p and q.

$\mathbf{F} = m\mathbf{a}$

$$\begin{pmatrix} 5 + 10p \\ 150 + 20q \end{pmatrix} = 60 \times \begin{pmatrix} 0.8 \\ -1.5 \end{pmatrix} = \begin{pmatrix} 48 \\ -90 \end{pmatrix}$$ ——— Use $\mathbf{F} = m\mathbf{a}$. Remember that you need to multiply each component in the acceleration by 60.

So $5 + 10p = 48 \implies p = 4.3$
and $150 + 20q = -90 \implies q = -12$ ——— Solve separate equations for each component to find the values of p and q.

Exercise 10D

In all the questions in this exercise $\mathbf{i}$ represents the unit vector due east, and $\mathbf{j}$ represents the unit vector due north.

1 A resultant force of $(\mathbf{i} + 4\mathbf{j})$ N acts upon a particle of mass 2 kg.
   a Find the acceleration of the particle in the form $(p\mathbf{i} + q\mathbf{j})$ m s^{-2}.
   b Find the magnitude and bearing of the acceleration of the particle.

2 A resultant force of $(4\mathbf{i} + 3\mathbf{j})$ N acts on a particle of mass m kg causing it to accelerate at $(20\mathbf{i} + 15\mathbf{j})$ m s^{-2}. Work out the mass of the particle.

3 A particle of mass 3 kg is acted on by a force $\mathbf{F}$. Given that the particle accelerates at $(7\mathbf{i} - 3\mathbf{j})\,\mathrm{m\,s^{-2}}$:

   a find an expression for $\mathbf{F}$ in the form $(p\mathbf{i} + q\mathbf{j})\,\mathrm{N}$

   b find the magnitude and bearing of $\mathbf{F}$.

4 Two forces, $\mathbf{F}_1$ and $\mathbf{F}_2$, act on a particle of mass m. Find the acceleration of the particle, $\mathbf{a}\,\mathrm{m\,s^{-2}}$, given that:

   a $\mathbf{F}_1 = (2\mathbf{i} + 7\mathbf{j})\,\mathrm{N}$, $\mathbf{F}_2 = (-3\mathbf{i} + \mathbf{j})\,\mathrm{N}$, $m = 0.25\,\mathrm{kg}$

   b $\mathbf{F}_1 = (3\mathbf{i} - 4\mathbf{j})\,\mathrm{N}$, $\mathbf{F}_2 = (2\mathbf{i} + 3\mathbf{j})\,\mathrm{N}$, $m = 6\,\mathrm{kg}$

   c $\mathbf{F}_1 = (-40\mathbf{i} - 20\mathbf{j})\,\mathrm{N}$, $\mathbf{F}_2 = (25\mathbf{i} + 10\mathbf{j})\,\mathrm{N}$, $m = 15\,\mathrm{kg}$

   d $\mathbf{F}_1 = 4\mathbf{j}\,\mathrm{N}$, $\mathbf{F}_2 = (-2\mathbf{i} + 5\mathbf{j})\,\mathrm{N}$, $m = 1.5\,\mathrm{kg}$

> **Notation** You are asked to find the acceleration as a vector, **a**. You can give your answer as a column vector or using **i–j** notation.

5 A particle of mass 8 kg is at rest. It is acted on by three forces, $\mathbf{F}_1 = \begin{pmatrix} 3 \\ -1 \end{pmatrix}\,\mathrm{N}$, $\mathbf{F}_2 = \begin{pmatrix} 2 \\ -5 \end{pmatrix}\,\mathrm{N}$ and $\mathbf{F}_3 = \begin{pmatrix} -1 \\ 0 \end{pmatrix}\,\mathrm{N}$.

   a Find the magnitude and direction of the acceleration of the particle, $\mathbf{a}\,\mathrm{m\,s^{-2}}$.

   b Find the time taken for the particle to travel a distance of 20 m.

> **Hint** Use $s = ut + \frac{1}{2}at^2$ with $s = 20$ and $u = 0$.

(E/P) 6 Two forces, $(2\mathbf{i} + 3\mathbf{j})\,\mathrm{N}$ and $(p\mathbf{i} + q\mathbf{j})\,\mathrm{N}$, act on a particle P. The resultant of the two forces is $\mathbf{R}$. Given that $\mathbf{R}$ acts in a direction which is parallel to the vector $(-\mathbf{i} + 4\mathbf{j})$, show that $4p + q + 11 = 0$. **(4 marks)**

> **Problem-solving** You can write **R** in the form $(-k\mathbf{i} + 4k\mathbf{j})\,\mathrm{N}$ for some constant k.

(E) 7 A particle of mass 4 kg starts from rest and is acted upon by a force $\mathbf{R}$ of $(6\mathbf{i} + b\mathbf{j})\,\mathrm{N}$. $\mathbf{R}$ acts on a bearing of 045°.

   a Find the value of b. **(1 mark)**

   b Calculate the magnitude of $\mathbf{R}$. **(2 marks)**

   c Work out the magnitude of the acceleration of the particle. **(2 marks)**

   d Find the total distance travelled by the particle during the first 5 seconds of its motion. **(3 marks)**

(P) 8 Three forces, $\mathbf{F}_1$, $\mathbf{F}_2$ and $\mathbf{F}_3$ act on a particle. $\mathbf{F}_1 = (-3\mathbf{i} + 7\mathbf{j})\,\mathrm{N}$, $\mathbf{F}_2 = (\mathbf{i} - \mathbf{j})\,\mathrm{N}$ and $\mathbf{F}_3 = (p\mathbf{i} + q\mathbf{j})\,\mathrm{N}$.

   a Given that this particle is in equilibrium, determine the value of p and the value of q.

   Force $\mathbf{F}_2$ is removed.

   b Given that in the first 10 seconds of its motion the particle travels a distance of 12 m, find the exact mass of the particle in kg.

(P) 9 A particle of mass m kg is acted upon by forces of $(5\mathbf{i} + 6\mathbf{j})\,\mathrm{N}$, $(2\mathbf{i} - 2\mathbf{j})\,\mathrm{N}$ and $(-\mathbf{i} - 4\mathbf{j})\,\mathrm{N}$ causing it to accelerate at $7\,\mathrm{m\,s^{-2}}$. Work out the mass of the particle. Give your answer correct to 2 d.p.

10 Two forces, $\binom{2}{5}$ N and $\binom{p}{q}$ N, act on a particle P of mass m kg. The resultant of the two forces is **R**.

   a Given that **R** acts in a direction which is parallel to the vector $\binom{1}{-2}$, show that
     $2p + q + 9 = 0$.   **(4 marks)**

   b Given also that $p = 1$ and that P moves with an acceleration of magnitude
     $15\sqrt{5}$ m s^{-2}, find the value of m.   **(7 marks)**

Challenge

A particle of mass 0.5 kg is acted on by two forces:

$\mathbf{F}_1 = -4\mathbf{i}$ N $\mathbf{F}_2 = (k\mathbf{i} + 2k\mathbf{j})$ N

where k is a positive constant.

Given that the particle is accelerating at a rate of $8\sqrt{17}$ m s^{-2}, find the value of k.

10.5 Connected particles

If a system involves the motion of more than one particle, the particles may be considered separately. However, if all parts of the system are moving in the **same straight line**, then you can also treat the whole system as a single particle.

- **You can solve problems involving connected particles by considering the particles separately or, if they are moving in the same straight line, as a single particle.**

> **Watch out**   Particles need to remain in contact, or be connected by an inextensible rod or string to be considered as a single particle.

Example 11

Two particles, P and Q, of masses 5 kg and 3 kg respectively, are connected by a light inextensible string. Particle P is pulled by a horizontal force of magnitude 40 N along a rough horizontal plane.

Particle P experiences a frictional force of 10 N and particle Q experiences a frictional force of 6 N.

a Find the acceleration of the particles.

b Find the tension in the string.

c Explain how the modelling assumptions that the string is light and inextensible have been used.

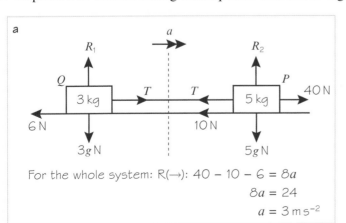

For the whole system: R(→): $40 - 10 - 6 = 8a$
$$8a = 24$$
$$a = 3 \text{ m s}^{-2}$$

> **Problem-solving**
>
> In part **a**, by considering the system as a single particle you eliminate the need to find the tension in the string. Otherwise you would need to set up two simultaneous equations involving a and T.
>
> In part **b** the particles need to be considered separately to find the tension in the string.

b For P: $R(\rightarrow)$: $40 - T - 10 = 5 \times 3$

$T = 15 \text{ N}$ •————

c Inextensible $\Rightarrow$ acceleration of masses is the same.

light $\Rightarrow$ tension is the same throughout the length of the string and the mass of the string is negligible.

You could also have chosen particle Q to find the tension. Check to see that it gives the same answer.

■ **Newton's third law states that for every action there is an equal and opposite reaction.**

Newton's third law means that when two bodies A and B are in contact, if body A exerts a force on body B, then body B exerts a force on body A that is equal in magnitude and acts in the opposite direction.

Example 12

A light scale-pan is attached to a vertical light inextensible string. The scale-pan carries two masses A and B. The mass of A is 400 g and the mass of B is 600 g. A rests on top of B, as shown in the diagram.

The scale-pan is raised vertically, using the string, with acceleration 0.5 m s^{-2}.

a Find the tension in the string.

b Find the force exerted on mass B by mass A.

c Find the force exerted on mass B by the scale-pan.

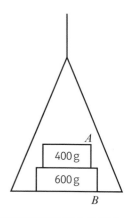

a For the whole system: •————

$R(\uparrow)$: $T - 0.4g - 0.6g = (0.4 + 0.6)a$

So, $T - g = 1 \times 0.5$ •————

$T = 10.3 \text{ N}$ •————

The tension in the string is 10 N (2 s.f.)

You can use this since all parts of the system are moving in the same straight line.

Note that you must convert 400 g to 0.4 kg and 600 g to 0.6 kg.

$a = 0.5$

Simplify.

b

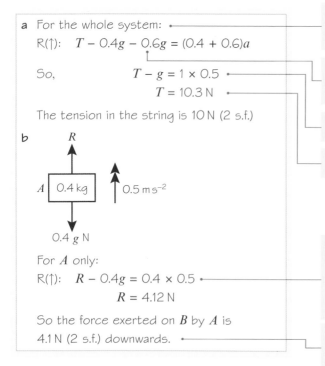

For A only:

$R(\uparrow)$: $R - 0.4g = 0.4 \times 0.5$ •————

$R = 4.12 \text{ N}$

So the force exerted on B by A is 4.1 N (2 s.f.) downwards. •————

Find the force exerted on A by B and then use Newton's 3rd law to say that the force exerted on B by A will have the same magnitude but is in the opposite direction.

You have used $g = 9.8 \text{ m s}^{-2}$ so give your final answer correct to two significant figures.

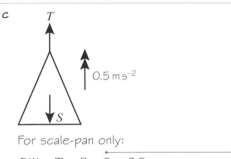

c

For scale-pan only:

$R(\uparrow) \quad T - S = 0 \times 0.5$

$\qquad\qquad = 0$

So, $\qquad T = S = 10.3\,\text{N}$

So, the force exerted on B by the scale-pan is 10 N (2 s.f.) upwards.

Exercise 10E

1 Two particles P and Q of masses 8 kg and 2 kg respectively, are connected by a light inextensible string. The particles are on a smooth horizontal plane. A horizontal force of magnitude F is applied to P in a direction away from Q and when the string is taut the particles move with acceleration $0.4\,\text{m s}^{-2}$.

  a Find the value of F.

  b Find the tension in the string.

  c Explain how the modelling assumptions that the string is light and inextensible are used.

2 Two particles P and Q of masses 20 kg and m kg are connected by a light inextensible rod. The particles lie on a smooth horizontal plane. A horizontal force of 60 N is applied to Q in a direction towards P, causing the particles to move with acceleration $2\,\text{m s}^{-2}$.

  a Find the mass, m, of Q.

  b Find the thrust in the rod.

> **Hint** For part **b** consider P on its own.

3 Two particles P and Q of masses 7 kg and 8 kg are connected by a light inextensible string. The particles are on a smooth horizontal plane. A horizontal force of 30 N is applied to Q in a direction away from P. When the string is taut the particles move with acceleration, $a\,\text{m s}^{-2}$.

  a Find the acceleration, a, of the system.

  b Find the tension in the string.

4 Two boxes A and B of masses 110 kg and 190 kg sit on the floor of a lift of mass 1700 kg. Box A rests on top of box B. The lift is supported by a light inextensible cable and is descending with constant acceleration $1.8\,\text{m s}^{-2}$.

  a Find the tension in the cable.

  b Find the force exerted by box B

    i on box A **ii** on the floor of the lift.

(P) **5** A lorry of mass m kg is towing a trailer of mass $3m$ kg along a straight horizontal road. The lorry and trailer are connected by a light inextensible tow-bar. The lorry exerts a driving force of 50 000 N causing the lorry and trailer to accelerate at $5\,\mathrm{m\,s^{-2}}$. The lorry and trailer experience resistances of 4000 N and 10 000 N respectively.

   a Find the mass of the lorry and hence the mass of the trailer.

   b Find the tension in the tow-bar.

   c Explain how the modelling assumptions that the tow-bar is light and inextensible affect your calculations.

(E) **6** Two particles A and B of masses 10 kg and 5 kg respectively are connected by a light inextensible string. Particle B hangs directly below particle A. A force of 180 N is applied vertically upwards causing the particles to accelerate.

   a Find the magnitude of the acceleration. **(3 marks)**

   b Find the tension in the string. **(2 marks)**

(E/P) **7** Two particles A and B of masses 6 kg and m kg respectively are connected by a light inextensible string. Particle B hangs directly below particle A. A force of 118 N is applied vertically upwards causing the particles to accelerate at $2\,\mathrm{m\,s^{-2}}$.

   a Find the mass, m, of particle B. **(3 marks)**

   b Find the tension in the string. **(2 marks)**

(E/P) **8** A train engine of mass 6400 kg is pulling a carriage of mass 1600 kg along a straight horizontal railway track. The engine is connected to the carriage by a shunt which is parallel to the direction of motion of the coupling. The shunt is modelled as a light rod. The engine provides a constant driving force of 12 000 N. The resistances to the motion of the engine and the carriage are modelled as constant forces of magnitude R N and 2000 N respectively.

   Given that the acceleration of the engine and the carriage is $0.5\,\mathrm{m\,s^{-2}}$:

   a find the value of R **(3 marks)**

   b show that the tension in the shunt is 2800 N. **(2 marks)**

(E) **9** A car of mass 900 kg pulls a trailer of mass 300 kg along a straight horizontal road using a light tow-bar which is parallel to the road. The horizontal resistances to motion of the car and the trailer have magnitudes 200 N and 100 N respectively. The engine of the car produces a constant horizontal driving force on the car of magnitude 1200 N.

   a Show that the acceleration of the car and trailer is $0.75\,\mathrm{m\,s^{-2}}$. **(2 marks)**

   b Find the magnitude of the tension in the tow-bar. **(3 marks)**

   The car is moving along the road when the driver sees a set of traffic lights have turned red. He reduces the force produced by the engine to zero and applies the brakes. The brakes produce a force on the car of magnitude F newtons and the car and trailer decelerate.

   c Given that the resistances to motion are unchanged and the magnitude of the thrust in the towbar is 100 N, find the value of F. **(7 marks)**

10.6 Pulleys

In this section you will see how to model systems of connected particles involving pulleys. The problems you answer will assume that particles are connected by a light, inextensible string, which passes over a **smooth pulley**. This means that the tension in the string will be the same **on both sides** of the pulley. The parts of the string each side of the pulley will be either horizontal or vertical.

Watch out You cannot treat a system involving a pulley as a single particle. This is because the particles are moving in different directions.

Example 13

Particles P and Q, of masses $2m$ and $3m$, are attached to the ends of a light inextensible string. The string passes over a small smooth fixed pulley and the masses hang with the string taut. The system is released from rest.

a i Write down an equation of motion for P.
   ii Write down an equation of motion for Q.
b Find the acceleration of each mass.
c Find the tension in the string.
d Find the force exerted on the pulley by the string.
e Find the distance moved by Q in the first 4 s, assuming that P does not reach the pulley.
f State how you have used the fact that the pulley is smooth in your calculations.

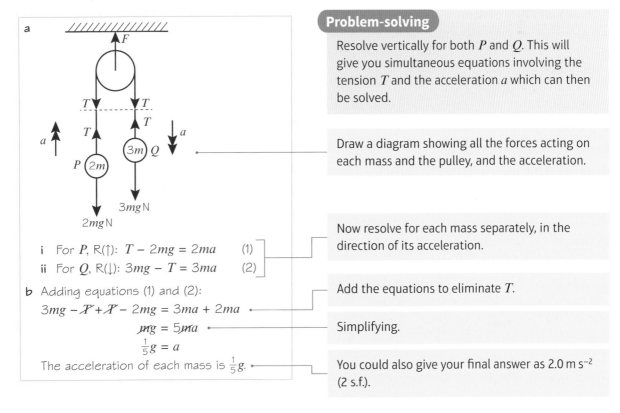

Problem-solving

Resolve vertically for both P and Q. This will give you simultaneous equations involving the tension T and the acceleration a which can then be solved.

Draw a diagram showing all the forces acting on each mass and the pulley, and the acceleration.

i For P, R(↑): $T - 2mg = 2ma$ (1)
ii For Q, R(↓): $3mg - T = 3ma$ (2)

Now resolve for each mass separately, in the direction of its acceleration.

b Adding equations (1) and (2):

$3mg - \cancel{T} + \cancel{T} - 2mg = 3ma + 2ma$

Add the equations to eliminate T.

$mg = 5ma$

Simplifying.

$\frac{1}{5}g = a$

The acceleration of each mass is $\frac{1}{5}g$.

You could also give your final answer as 2.0 m s^{-2} (2 s.f.).

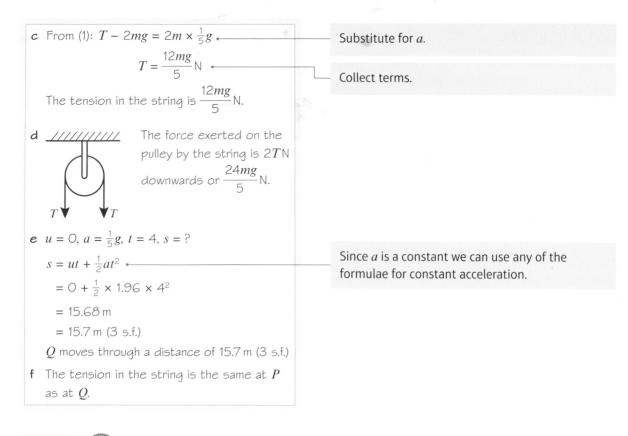

c From (1): $T - 2mg = 2m \times \frac{1}{5}g$ ──── Substitute for a.

$$T = \frac{12mg}{5} N$$ ──── Collect terms.

The tension in the string is $\frac{12mg}{5}$ N.

d The force exerted on the pulley by the string is $2T$ N downwards or $\frac{24mg}{5}$ N.

e $u = 0$, $a = \frac{1}{5}g$, $t = 4$, $s = ?$

$s = ut + \frac{1}{2}at^2$ ──── Since a is a constant we can use any of the formulae for constant acceleration.

$= 0 + \frac{1}{2} \times 1.96 \times 4^2$

$= 15.68$ m

$= 15.7$ m (3 s.f.)

Q moves through a distance of 15.7 m (3 s.f.)

f The tension in the string is the same at P as at Q.

Example 14

Two particles A and B of masses 0.4 kg and 0.8 kg respectively are connected by a light inextensible string. Particle A lies on a rough horizontal table 4.5 m from a small smooth pulley which is fixed at the edge of the table. The string passes over the pulley and B hangs freely, with the string taut, 0.5 m above horizontal ground. A frictional force of magnitude $0.08g$ opposes the motion of particle A. The system is released from rest. Find:

a the acceleration of the system

b the time taken for B to reach the ground

c the total distance travelled by A before it first comes to rest.

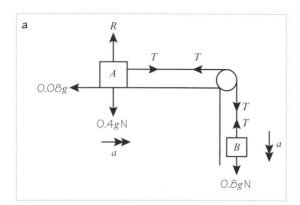

Problem-solving

Draw a diagram showing all the forces and the accelerations. The pulley is smooth so the tension in the string is the same on each side of the pulley.

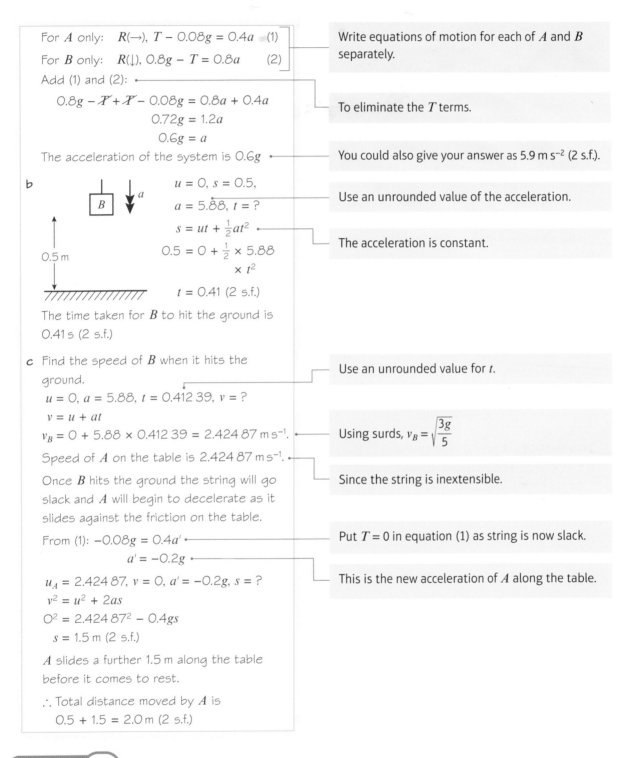

For A only: $R(\rightarrow)$, $T - 0.08g = 0.4a$ (1)

For B only: $R(\downarrow)$, $0.8g - T = 0.8a$ (2)

> Write equations of motion for each of A and B separately.

Add (1) and (2):

$$0.8g - \cancel{T} + \cancel{T} - 0.08g = 0.8a + 0.4a$$
$$0.72g = 1.2a$$
$$0.6g = a$$

> To eliminate the T terms.

The acceleration of the system is $0.6g$

> You could also give your answer as $5.9\ \text{m s}^{-2}$ (2 s.f.).

b

$u = 0$, $s = 0.5$,
$a = 5.\dot{8}\dot{8}$, $t = ?$

> Use an unrounded value of the acceleration.

$$s = ut + \tfrac{1}{2}at^2$$
$$0.5 = 0 + \tfrac{1}{2} \times 5.88 \times t^2$$

> The acceleration is constant.

0.5 m

$$t = 0.41 \ (2 \text{ s.f.})$$

The time taken for B to hit the ground is $0.41\ \text{s}$ (2 s.f.)

c Find the speed of B when it hits the ground.

> Use an unrounded value for t.

$u = 0$, $a = 5.88$, $t = 0.412\,39$, $v = ?$

$$v = u + at$$
$$v_B = 0 + 5.88 \times 0.412\,39 = 2.424\,87\ \text{m s}^{-1}.$$

> Using surds, $v_B = \sqrt{\dfrac{3g}{5}}$

Speed of A on the table is $2.424\,87\ \text{m s}^{-1}$.

Once B hits the ground the string will go slack and A will begin to decelerate as it slides against the friction on the table.

> Since the string is inextensible.

From (1): $-0.08g = 0.4a'$

> Put $T = 0$ in equation (1) as string is now slack.

$$a' = -0.2g$$

> This is the new acceleration of A along the table.

$u_A = 2.424\,87$, $v = 0$, $a' = -0.2g$, $s = ?$

$$v^2 = u^2 + 2as$$
$$0^2 = 2.424\,87^2 - 0.4gs$$
$$s = 1.5\ \text{m} \ (2 \text{ s.f.})$$

A slides a further $1.5\ \text{m}$ along the table before it comes to rest.

$\therefore$ Total distance moved by A is
$$0.5 + 1.5 = 2.0\ \text{m} \ (2 \text{ s.f.})$$

Exercise 10F

Ⓟ **1** Two particles A and B of masses $4\ \text{kg}$ and $3\ \text{kg}$ respectively are connected by a light inextensible string which passes over a small smooth fixed pulley. The particles are released from rest with the string taut.

  a Find the tension in the string.

When A has travelled a distance of 2 m it strikes the ground and immediately comes to rest.

b Find the speed of A when it hits the ground.

c Assuming that B does not hit the pulley, find the greatest height that B reaches above its initial position.

> **Problem-solving**
>
> After A hits the ground B behaves like a particle moving freely under gravity.

E/P **2** Two particles P and Q have masses km and $3m$ respectively, where $k < 3$. The particles are connected by a light inextensible string which passes over a smooth light fixed pulley. The system is held at rest with the string taut, the hanging parts of the string vertical and with P and Q at the same height above a horizontal plane, as shown in the diagram. The system is released from rest. After release, Q descends with acceleration $\frac{1}{3}g$.

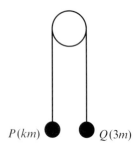

a Calculate the tension in the string as Q descends. **(3 marks)**

b Show that $k = 1.5$ **(3 marks)**

c State how you have used the information that the pulley is smooth. **(1 mark)**

After descending for 1.8 s, the particle Q reaches the plane. It is immediately brought to rest by the impact with the plane. The initial distance between P and the pulley is such that, in the subsequent motion, P does not reach the pulley.

d Show that the greatest height, in metres, reached by P above the plane is $1.26\,g$. **(7 marks)**

E/P **3** Two particles A and B have masses m kg and 3 kg respectively, where $m > 3$. The particles are connected by a light inextensible string which passes over a smooth, fixed pulley. Initially A is 2.5 m above horizontal ground. The particles are released from rest with the string taut and the hanging parts of the string vertical, as shown in the figure. After A has been descending for 1.25 s, it strikes the ground. Particle A reaches the ground before B has reached the pulley.

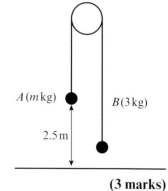

a Show that the acceleration of B as it ascends is $3.2\,\text{m s}^{-2}$. **(3 marks)**

b Find the tension in the string as A descends. **(3 marks)**

c Show that $m = \frac{65}{11}$. **(4 marks)**

d State how you have used the information that the string is inextensible. **(1 mark)**

When A strikes the ground it does not rebound and the string becomes slack. Particle B then moves freely under gravity, without reaching the pulley, until the string becomes taut again.

e Find the time between the instant when A strikes the ground and the instant when the string becomes taut again. **(6 marks)**

4 Two particles A and B of masses 5 kg and 3 kg respectively are connected by a light inextensible string. Particle A lies on a rough horizontal table and the string passes over a small smooth pulley which is fixed at the edge of the table. Particle B hangs freely. The friction between A and the table is 24.5 N. The system is released from rest. Find:

a the acceleration of the system

b the tension in the string

c the magnitude of the force exerted on the pulley by the string.

(E) 5 A box P of mass 2.5 kg rests on a rough
horizontal table and is attached to one end of
a light inextensible string. The string passes
over a small smooth pulley fixed at the edge
of the table. The other end of the string is
attached to a sphere Q of mass 1.5 kg which
hangs freely below the pulley. The magnitude
of the frictional force between P and the table
is k N. The system is released from rest with
the string taut. After release, Q descends a distance of 0.8 m in 0.75 s.

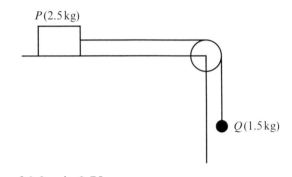

a Modelling P and Q as particles:

  i   calculate the acceleration of Q **(3 marks)**

  ii show that the tension in the string is 10.4 N (to 3 s.f.)   **(4 marks)**

  iii find the value of k.   **(3 marks)**

b State how in your calculations you have used the information that the string is
   inextensible.   **(1 mark)**

Mixed exercise 10

1 A motorcycle of mass 200 kg is moving along a level road. The motorcycle's engine provides
a forward thrust of 1000 N. The total resistance is modelled as a constant force of magnitude
600 N.

a Modelling the motorcycle as a particle, draw a force diagram to show the forces acting on the
   motorcycle.

b Calculate the acceleration of the motorcycle.

2 A man of mass 86 kg is standing in a lift which is moving upwards with constant acceleration
2 m s^{-2}. Find the magnitude and direction of the force that the man is exerting on the floor of
the lift.

(P) 3 A car of mass 800 kg is travelling along a straight horizontal road. A constant retarding force
of F N reduces the speed of the car from 18 m s^{-1} to 12 m s^{-1} in 2.4 s. Calculate:

a the value of F b the distance moved by the car in these 2.4 s.

(E) 4 A block of mass 0.8 kg is pushed along a rough horizontal floor by a constant horizontal force
of magnitude 7 N. The speed of the block increases from 2 m s^{-1} to 4 m s^{-1} in a distance of
4.8 m. Calculate:

a the magnitude of the acceleration of the block   **(3 marks)**

b the magnitude of the frictional force between the block and the floor. **(3 marks)**

(P) 5 A car of mass 1200 kg is moving along a level road. The car's engine provides a constant driving force. The motion of the car is opposed by a constant resistance.

Given that car is accelerating at $2\,\mathrm{m\,s^{-2}}$, and that the magnitude of the driving force is three times the magnitude of the resistance force, show that the magnitude of the driving force is 3600 N.

6 Forces of $(3\mathbf{i} + 2\mathbf{j})$ N and $(4\mathbf{i} - \mathbf{j})$ N act on a particle of mass 0.25 kg. Find the acceleration of the particle.

(P) 7 Forces of $\begin{pmatrix} 2 \\ -1 \end{pmatrix}$ N, $\begin{pmatrix} 3 \\ -1 \end{pmatrix}$ N and $\begin{pmatrix} a \\ -2b \end{pmatrix}$ N act on a particle of mass 2 kg causing it to accelerate

at $\begin{pmatrix} 3 \\ 2 \end{pmatrix}$ m s^{-2}. Find the values of a and b.

8 A sled of mass 2 kg is initially at rest on a horizontal plane. It is acted upon by a force of $(2\mathbf{i} + 4\mathbf{j})$ N for 3 seconds. Giving your answers in surd form,

    a find the magnitude of acceleration

    b find the distance travelled in the 3 seconds.

(P) 9 In this question $\mathbf{i}$ and $\mathbf{j}$ represent the unit vectors east and north respectively.

The forces $(3a\mathbf{i} + 4b\mathbf{j})$ N, $(5b\mathbf{i} + 2a\mathbf{j})$ N and $(-15\mathbf{i} - 18\mathbf{j})$ N act on a particle of mass 2 kg which is in equilibrium.

    a Find the values of a and b.

    b The force $(-15\mathbf{i} - 18\mathbf{j})$ N is removed. Work out:

        i the magnitude and direction of the resulting acceleration of the particle

        ii the distance travelled by the particle in the first 3 seconds of its motion.

(E/P) 10 A car is towing a trailer along a straight horizontal road by means of a horizontal tow-rope. The mass of the car is 1400 kg. The mass of the trailer is 700 kg. The car and the trailer are modelled as particles and the tow-rope as a light inextensible string. The resistances to motion of the car and the trailer are assumed to be constant and of magnitude 630 N and 280 N respectively. The driving force on the car, due to its engine, is 2380 N. Find:

    a the acceleration of the car   **(3 marks)**

    b the tension in the tow-rope.   **(3 marks)**

When the car and trailer are moving at $12\,\mathrm{m\,s^{-1}}$, the tow-rope breaks. Assuming that the driving force on the car and the resistances to motion are unchanged:

    c find the distance moved by the car in the first 4 s after the tow-rope breaks.   **(6 marks)**

    d State how you have used the modelling assumption that the tow-rope is inextensible.   **(1 mark)**

(E/P) 11 A train of mass 2500 kg pushes a carriage of mass 1100 kg along a straight horizontal track. The engine is connected to the carriage by a shunt which is parallel to the direction of motion of the coupling. The horizontal resistances to motion of the train and the carriage have magnitudes R N and 500 N respectively. The engine of the train produces a constant horizontal driving force of magnitude 8000 N that causes the train and carriage to accelerate at $1.75\,\mathrm{m\,s^{-2}}$.

    a Show that the resistance to motion R acting on the train is 1200 N.   **(2 marks)**

    b Find the magnitude of the compression in the shunt.   **(3 marks)**

The train must stop at the next station so the driver reduces the force produced by the engine to zero and applies the brakes. The brakes produce a force on the train of magnitude 2000 N causing the engine and carriage to decelerate.

    c Given that the resistances to motion are unchanged, find the magnitude of the thrust in the shunt. Give your answer correct to 3 s.f.   **(7 marks)**

(P) **12** Particles P and Q of masses $2m$ kg and m kg respectively are attached to the ends of a light inextensible string which passes over a smooth fixed pulley. They both hang at a distance of 2 m above horizontal ground. The system is released from rest.

    a Find the magnitude of the acceleration of the system.

    b Find the speed of P as it hits the ground.

Given that particle Q does not reach the pulley:

    c find the greatest height that Q reaches above the ground.

    d State how you have used in your calculation:

       i the fact that the string is inextensible   **ii** the fact that the pulley is smooth.

(E/P) **13** Two particles have masses 3 kg and m kg, where $m < 3$. They are attached to the ends of a light inextensible string. The string passes over a smooth fixed pulley. The particles are held in position with the string taut and the hanging parts of the string vertical, as shown. The particles are then released from rest. The initial acceleration of each particle has magnitude $\frac{3}{7}g$. Find:

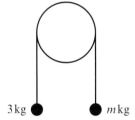

3 kg m kg

    a the tension in the string immediately after the particles are released   **(3 marks)**

    b the value of m.   **(3 marks)**

(E/P) **14** A block of wood A of mass 0.5 kg rests on a rough horizontal table and is attached to one end of a light inextensible string. The string passes over a small smooth pulley P fixed at the edge of the table. The other end of the string is attached to a ball B of mass 0.8 kg which hangs freely below the pulley, as shown in the figure. The resistance to motion of A from the rough table has a constant magnitude

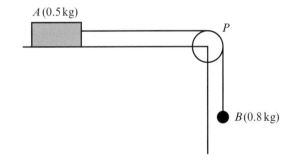

A (0.5 kg)

P

B (0.8 kg)

F N. The system is released from rest with the string taut. After release, B descends a distance of 0.4 m in 0.5 s. Modelling A and B as particles, calculate:

    a the acceleration of B **(3 marks)**

    b the tension in the string   **(4 marks)**

    c the value of F.   **(3 marks)**

    d State how in your calculations you have used the information that the string is inextensible.   **(1 mark)**

15 Two particles P and Q have masses $0.5\,\text{kg}$ and $0.4\,\text{kg}$ respectively. The particles are attached to the ends of a light inextensible string. The string passes over a small smooth pulley which is fixed above a horizontal floor. Both particles are held, with the string taut, at a height of $2\,\text{m}$ above the floor, as shown. The particles are released from rest and in the subsequent motion Q does not reach the pulley.

a i Write down an equation of motion for P. **(2 marks)**

ii Write down an equation of motion for Q. **(2 marks)**

b Find the tension in the string immediately after the particles are released.
 (2 marks)

c Find the acceleration of A immediately after the particles are released. **(2 marks)**

When the particles have been moving for $0.2\,\text{s}$, the string breaks.

d Find the further time that elapses until Q hits the floor. **(9 marks)**

Challenge

In this question **i** and **j** are the unit vectors east and north respectively.

Two boats start from rest at different points on the south bank of a river. The current in the river provides a constant force of magnitude $3\mathbf{i}$ N on both boats.

The motor on boat A provides a thrust of $(-7\mathbf{i} + 2\mathbf{j})$ N and the motor on boat B provides a thrust of $(k\mathbf{i} + \mathbf{j})$ N.

Given that the boats are accelerating in perpendicular directions, find the value of k.

Summary of key points

1 **Newton's first law** of motion states that an object at rest will stay at rest and that an object moving with constant velocity will continue to move with constant velocity unless an unbalanced force acts on the object.

2 A **resultant** force acting on an object will cause the object to **accelerate in the same direction** as the resultant force.

3 You can find the **resultant** of two or more forces given as vectors by adding the vectors.

4 **Newton's second law** of motion states that the force needed to accelerate a particle is equal to the product of the mass of the particle and the acceleration produced: $F = ma$.

5 $W = mg$

6 You can use $\mathbf{F} = m\mathbf{a}$ to solve problems involving vector forces acting on particles.

7 You can solve problems involving connected particles by considering the particles separately or, if they are moving in the same straight line, as a single particle.

8 **Newton's third law** states that for every action there is an equal and opposite reaction.

Variable acceleration

Objectives

After completing this chapter you should be able to:

* Understand that displacement, velocity and acceleration may be given as functions of time → **pages 181–184**
* Use differentiation to solve kinematics problems → **pages 185–186**
* Use calculus to solve problems involving maxima and minima → **pages 186–188**
* Use integration to solve kinematics problems → **pages 188–191**
* Use calculus to derive constant acceleration formulae → **pages 191–193**

A space rocket experiences **variable acceleration** during launch. The **rate of change of velocity** increases to enable the rocket to escape the gravitational pull of the Earth.

→ **Mixed exercise Q13**

Prior knowledge check

1. Find $\dfrac{dy}{dx}$ given:

   a $y = 3x^2 - 5x + 6$ **b** $y = 2\sqrt{x} + \dfrac{6}{x^2} - 1$

   ← **Pure Year 1, Chapter 12**

2. Find the coordinates of the turning points on the curve with equation:

   a $y = 3x^2 - 9x + 2$ **b** $y = x^3 - 6x^2 + 9x + 5$

   ← **Pure Year 1, Chapter 12**

3. Find f(x) given:

   a f$'(x) = 5x + 8$, f$(0) = 1$
   b f$'(x) = 3x^2 - 2x + 5$, f$(0) = 7$

   ← **Pure Year 1, Chapter 13**

4. Find the area bounded by the x-axis and:

   a the line $y = 2x - 1$, $x = 2$ and $x = 5$
   b the curve $y = 6x - 2 - x^2$, $x = 1$ and $x = 3$

   ← **Pure Year 1, Chapter 13**

11.1 Functions of time

If acceleration of a moving particle is variable, it changes with time and can be expressed as a function of time.

In the same way, velocity and displacement can also be expressed as functions of time.

> **Links** Acceleration is the gradient of a velocity–time graph. ← **Section 9.2**

These velocity–time graphs represent the motion of a particle travelling in a straight line. They show examples of increasing and decreasing acceleration.

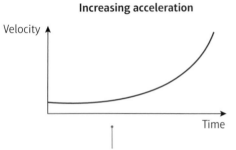

Increasing acceleration

The rate of increase of velocity is increasing with time and the gradient of the curve is increasing.

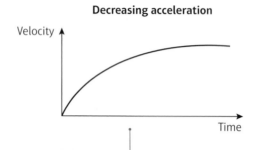

Decreasing acceleration

The rate of increase of velocity is decreasing with time and the gradient of the curve is decreasing.

Example 1

A body moves in a straight line, such that its displacement, s metres, from a point O at time t seconds is given by $s = 2t^3 - 3t$ for $t > 0$. Find:

a s when $t = 2$ **b** the time taken for the particle to return to O.

a $s = 2 \times 2^3 - 3 \times 2$ Substitute $t = 2$ into the equation for s.
     $= 16 - 6 = 10$ metres

b $2t^3 - 3t = 0$ When the particle returns to the starting point, the displacement is equal to zero.
     $t(2t^2 - 3) = 0$ either $t = 0$ or $2t^2 = 3$

     $\Rightarrow t^2 = \dfrac{3}{2}$ so $t = \pm\sqrt{\dfrac{3}{2}}$ seconds

     Time taken to return to $O = \sqrt{\dfrac{3}{2}}$ seconds   Answer is $+\sqrt{\dfrac{3}{2}}$ as equation is only valid for $t > 0$.

Example 2

A toy train travels along a straight track, leaving the start of the track at time $t = 0$. It then returns to the start of the track. The distance, s metres, from the start of the track at time t seconds is modelled by:

$s = 4t^2 - t^3$, where $0 \leqslant t \leqslant 4$

Explain the restriction $0 \leqslant t \leqslant 4$.

s is distance from start of track so $s \geqslant 0$. ———— Use the initial conditions given.

So $4t^2 - t^3 \geqslant 0$

$t^2(4 - t) \geqslant 0$ ———— Distance is a scalar quantity and must be $\geqslant 0$.

$t^2 \geqslant 0$ for all t, and $(4 - t) < 0$ for all $t > 4$,

so $t^2(4 - t)$ is only non-negative for $t \leqslant 4$

Motion begins at $t = 0$ hence $t \geqslant 0$ ———— The restriction $t \geqslant 0$ is due to the motion beginning at $t = 0$, not due to the function.

Hence $0 \leqslant t \leqslant 4$

Problem-solving

You could also sketch the graph of $s = 4t^2 - t^3$ to show the values of t for which the model is valid.

Example (3)

A body moves in a straight line such that its velocity, $v\,\text{m s}^{-1}$, at time t seconds is given by:

$v = 2t^2 - 16t + 24$, for $t \geqslant 0$

Find:

a the initial velocity

b the values of t when the body is instantaneously at rest

c the value of t when the velocity is $64\,\text{m s}^{-1}$

d the greatest speed of the body in the interval $0 \leqslant t \leqslant 5$.

a $v = 0 - 0 + 24$ ———— The initial velocity means the velocity at $t = 0$.

$v = 24\,\text{m s}^{-1}$

b $2t^2 - 16t + 24 = 0$ ———— The body is at rest when $v = 0$, so solve the quadratic equation when $v = 0$.

$t^2 - 8t + 12 = 0$

$(t - 6)(t - 2) = 0$

Body at rest when $t = 2$ seconds and $t = 6$ seconds

c $2t^2 - 16t + 24 = 64$

$2t^2 - 16t - 40 = 0$ ———— Rearrange the quadratic equation to make it equal to zero and factorise.

$t^2 - 8t - 20 = 0$

$(t - 10)(t + 2) = 0$

Either $t = 10$ or $t = -2$

Velocity $= 64\,\text{m s}^{-1}$ when $t = 10$ seconds ———— The equation for velocity is valid for $t \geqslant 0$, so $t = -2$ is not a valid solution.

d

Sketch a velocity–time graph for the motion of the body. You can use the symmetry of the quadratic curve to determine the position of the turning point. ← **Pure Year 1, Chapter 2**

Watch out You need to find the greatest **speed**. This could occur when the velocity is positive or negative, so find the range of values taken by v in the interval $0 \leqslant t \leqslant 5$.

When $t = 4$, $v = 2(4)^2 - 16(4) + 24 = -8$

So in $0 \leqslant t \leqslant 5$ range of v is $-8 \leqslant v \leqslant 24$

Greatest speed is $24\,\text{m s}^{-1}$

Online Explore the solution using technology.

183

Exercise 11A

1 A body moves in a straight line such that its displacement, s metres, at time t seconds is given by $s = 9t - t^3$. Find:

 a s when $t = 1$ **b** the values of t when $s = 0$.

2 A particle P moves on the x-axis. At time t seconds the displacement s metres is given by $s = 5t^2 - t^3$. Find:

 a the change in displacement between $t = 2$ and $t = 4$

 b the change in displacement in the third second.

 > **Hint** The third second is the time between $t = 2$ and $t = 3$.

3 A particle moves in a straight line such that its velocity, $v\,\text{m s}^{-1}$, at time t seconds is given by $v = 3 + 5t - t^2$ for $t \geqslant 0$. Find:

 a the velocity of the particle when $t = 1$

 b the greatest speed of the particle in the interval $0 \leqslant t \leqslant 4$

 c the velocity of the particle when $t = 7$ and describe the direction of motion of the particle at this time.

4 At time $t = 0$, a toy car is at point P. It moves in a straight line from point P and then returns to P. Its distance from P, $s\,\text{m}$, at time t seconds can be modelled by $s = \frac{1}{5}(4t - t^2)$. Find:

 a the maximum displacement   **b** the time taken for the toy car to return to P

 c the total distance travelled   **d** the values of t for which the model is valid.

5 A body moves in a straight line such that its velocity, $v\,\text{m s}^{-1}$, at time t seconds is given by $v = 3t^2 - 10t + 8$, for $t \geqslant 0$. Find:

 a the initial velocity

 b the values of t when the body is instantaneously at rest

 c the values of t when the velocity is $5\,\text{m s}^{-1}$

 d the greatest speed of the body in the interval $0 \leqslant t \leqslant 2$.

E 6 A particle P moves on the x-axis. At time t seconds the velocity of P is $v\,\text{m s}^{-1}$ in the direction of x increasing, where $v = 8t - 2t^2$. When $t = 0$, P is at the origin O. Find:

 a the time taken for the particle to come to instantaneous rest   **(2 marks)**

 b the greatest speed of the particle in the interval $0 \leqslant t \leqslant 4$.   **(3 marks)**

E 7 At time $t = 0$, a particle moves in a straight horizontal line from a point O, then returns to the starting point. The distance, s metres, from the point O at time t seconds is given by:

 $$s = 3t^2 - t^3, \quad 0 \leqslant t \leqslant T$$

 Given that the model is valid when $s \geqslant 0$, find the value of T. Explain your answer.   **(3 marks)**

E 8 A particle P moves on the x-axis. At time t seconds the velocity of P is $v\,\text{m s}^{-1}$ in the direction of x increasing, where:

 $$v = \tfrac{1}{5}(3t^2 - 10t + 3), \qquad x \geqslant 0$$

 a Find the values of t when P is instantaneously at rest.   **(3 marks)**

 b Determine the greatest speed of P in the interval $0 \leqslant t \leqslant 3$.   **(4 marks)**

11.2 Using differentiation

Velocity is the rate of change of displacement.

- **If the displacement, s, is expressed as a function of t,**

  then the velocity, v, can be expressed as $v = \dfrac{ds}{dt}$

  Links The gradient of a displacement–time graph represents the velocity. ← Section 9.1

In the same way, acceleration is the rate of change of velocity.

- **If the velocity, v, is expressed as a function of t, then the acceleration, a, can be expressed as** $a = \dfrac{dv}{dt} = \dfrac{d^2s}{dt^2}$

  Links The gradient of a velocity–time graph represents the acceleration. ← Section 9.2

  $\dfrac{d^2s}{dt^2}$ is the second derivative (or second-order derivative) of s with respect to t.

  ← Pure Year 1, Chapter 12

Example 4

A particle P is moving on the x-axis. At time t seconds, the displacement x metres from O is given by $x = t^4 - 32t + 12$. Find:

a the velocity of P when $t = 3$

b the value of t for which P is instantaneously at rest

c the acceleration of P when $t = 1.5$.

a $\quad x = t^4 - 32t + 12$

$\quad v = \dfrac{dx}{dt} = 4t^3 - 32$

When $t = 3$,

$\quad v = 4 \times 3^3 - 32 = 76$

The velocity of P when $t = 3$ is $76\,\text{m s}^{-1}$ in the direction of x increasing.

You find the velocity by differentiating the displacement.

To find the velocity when $t = 3$, you substitute $t = 3$ into the expression.

b $\quad v = 4t^3 - 32 = 0$

$\quad t^3 = \dfrac{32}{4} = 8$

$\quad t = 2$

The particle is at rest when $v = 0$. You substitute $v = 0$ into your expression for v and solve the resulting equation to find t.

c $\quad v = 4t^3 - 32$

$\quad a = \dfrac{dv}{dt} = 12t^2$

When $t = 1.5$,

$\quad a = 12 \times 1.5^2 = 27$

The acceleration of P when $t = 1.5$ is $27\,\text{m s}^{-2}$.

You find the acceleration by differentiating the velocity.

Exercise 11B

1 Find an expression for **i** the velocity and **ii** the acceleration of a particle given that the displacement is given by:

a $s = 4t^4 - \dfrac{1}{t}$

b $x = \dfrac{2}{3}t^3 + \dfrac{1}{t^2}$

c $s = (3t^2 - 1)(2t + 5)$

d $x = \dfrac{3t^4 - 2t^3 + 5}{2t}$

2 A particle is moving in a straight line. At time t seconds, its displacement, x m, from a fixed point O on the line is given by $x = 2t^3 - 8t$. Find:

    a the velocity of the particle when $t = 3$ **b** the acceleration of the particle when $t = 2$.

(P) 3 A particle P is moving on the x-axis. At time t seconds (where $t \geqslant 0$), the velocity of P is v m s^{-1} in the direction of x increasing, where $v = 12 - t - t^2$.

    Find the acceleration of P when P is instantaneously at rest.

(P) 4 A particle is moving in a straight line. At time t seconds, its displacement, x m, from a fixed point O on the line is given by $x = 4t^3 - 39t^2 + 120t$.

    Find the distance between the two points where P is instantaneously at rest.

(E/P) 5 A particle P moves in a straight line. At time t seconds the acceleration of P is a m s^{-2} and the velocity v m s^{-1} is given by $v = kt - 3t^2$, where k is a constant.

    The initial acceleration of P is 4 m s^{-2}.

    a Find the value of k.   **(3 marks)**

    b Using the value of k found in part **a**, find the acceleration when P is instantaneously at rest.

                                                                                        (3 marks)

(E/P) 6 The print head on a printer moves such that its displacement s cm from the side of the printer at time t seconds is given by:

$$\tfrac{1}{4}(4t^3 - 15t^2 + 12t + 30), 0 \leqslant t \leqslant 3$$

    Find the distance between the points when the print head is instantaneously at rest, in cm to 1 decimal place.   **(6 marks)**

11.3 Maxima and minima problems

You can use calculus to determine maximum and minimum values of displacement, velocity and acceleration.

Example 5

A child is playing with a yo-yo. The yo-yo leaves the child's hand at time $t = 0$ and travels vertically in a straight line before returning to the child's hand. The distance, s m, of the yo-yo from the child's hand after time t seconds is given by:

$$s = 0.6t + 0.4t^2 - 0.2t^3, 0 \leqslant t \leqslant 3$$

a Justify the restriction $0 \leqslant t \leqslant 3$.

b Find the maximum distance of the yo-yo from the child's hand, correct to 3 s.f.

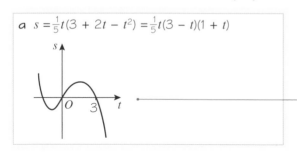

a $s = \tfrac{1}{5}t(3 + 2t - t^2) = \tfrac{1}{5}t(3 - t)(1 + t)$

> s is a cubic function with a negative coefficient of t^3, and roots at $t = -1$, $t = 0$ and $t = 3$.
>
> ← **Pure Year 1, Section 4.1**

$s = 0$ at $t = 0$ and $t = 3$

s is positive for all values of t between 0 and 3.

$s < 0$ for $t > 3$. Since s is a distance the model is not valid for $t > 3$.

> Comment on the value of s at the limits of the range **and** the behaviour of s within the range.

b $\dfrac{ds}{dt} = 0.6 + 0.8t - 0.6t^2$

$$\dfrac{ds}{dt} = 0$$

> You know from your answer to part **a** that the maximum value of s must occur at the turning point. To find the turning point differentiate and set $\dfrac{ds}{dt} = 0$.　　　　← **Pure Year 1, Section 12.9**

$0.6 + 0.8t - 0.6t^2 = 0$

$3t^2 - 4t - 3 = 0$

> Multiply each term by -5 to obtain an equation with a positive t^2 term and integer coefficients. This makes your working easier.

$t = \dfrac{4 \pm \sqrt{52}}{6} = 1.8685\ldots$ or $-0.5351\ldots$

$s = 0.6(1.8685\ldots) + 0.4(1.8685\ldots)^2$
$\quad - 0.2(1.8685\ldots)^3$
$\quad = 1.2129\ldots = 1.21\,\text{m (3 s.f.)}$

> Use the quadratic formula to solve the equation and take the positive value of t.

> Substitute this value of t back into the original equation to find the corresponding value of s. Remember to use unrounded values in your calculation, and check that your answer makes sense in the context of the question.

Exercise 11C

1 A particle P moves in a straight line such that its distance, s m, from a fixed point O at time t is given by:

$$s = 0.4t^3 - 0.3t^2 - 1.8t + 5,\ 0 \le t \le 3$$

The diagram shows the displacement–time graph of the motion of P.

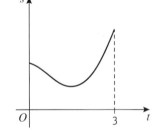

a Determine the time at which P is moving with minimum velocity.

b Find the displacement of P from O at this time.

c Find the velocity of P at this time.

P **2** A body starts at rest and moves in a straight line. At time t seconds the displacement of the body from its starting point, s m, is given by:

$$s = 4t^3 - t^4,\ 0 \le t \le 4.$$

a Show that the body returns to its starting position at $t = 4$.

b Explain why s is always non-negative.

c Find the maximum displacement of the body from its starting point.

> **Hint** Write $s = t^3(4 - t)$ and consider the sign of each factor in the range $0 \le t \le 4$.

3 At time $t = 0$ a particle P leaves the origin O and moves along the x-axis. At time t seconds the velocity of P is v m s^{-1}, where:

$$v = t^2(6 - t)^2,\ t \ge 0$$

a Sketch a velocity–time graph for the motion of P.

b Find the maximum value of v and the time at which it occurs.

(P) **4** A particle P moves along the x-axis. Its velocity, v m s^{-1} in the positive x-direction, at time t seconds is given by:

$$v = 2t^2 - 3t + 5, \, t \geq 0$$

    a Show that P never comes to rest.

    b Find the minimum velocity of P.

(E/P) **5** A particle P starts at the origin O at time $t = 0$ and moves along the x-axis. At time t seconds the distance of the particle, s m, from the origin is given by:

$$s = \frac{9t^2}{2} - t^3, \, 0 \leq t \leq 4.5$$

    a Sketch a displacement–time graph for the motion of P. **(2 marks)**

    b Hence justify the restriction $0 \leq t \leq 4.5$. **(2 marks)**

    c Find the maximum distance of the particle from O. **(5 marks)**

    d Find the magnitude of the acceleration of the particle at this point. **(3 marks)**

(E/P) **6** A train moves in a straight line along a 4 km test track. The motion of the train is modelled as a particle travelling in a straight line, and the distance, s m, of the train from the start of the track after time t seconds is given by $s = 3.6t + 1.76t^2 - 0.02t^3, \, 0 \leq t \leq 90$. Show that the train never reaches the end of the track. **(7 marks)**

11.4 Using integration

Integration is the reverse process to differentiation. You can integrate acceleration with respect to time to find velocity, and you can integrate velocity with respect to time to find displacement.

> **Links** The area under a velocity–time graph represents the displacement. ← **Section 9.2**

■

Differentiate   displacement $\quad = s = \int v \, dt$ ↑

$\dfrac{ds}{dt} = $ **velocity** $\quad = v = \int a \, dt$ **Integrate**

$\dfrac{dv}{dt} = \dfrac{d^2s}{dt^2} = $ **acceleration** $= a$

Example 6

A particle is moving on the x-axis. At time $t = 0$, the particle is at the point where $x = 5$. The velocity of the particle at time t seconds (where $t \geq 0$) is $(6t - t^2)$ m s^{-1}. Find:

a an expression for the displacement of the particle from O at time t seconds

b the distance of the particle from its starting point when $t = 6$.

a $x = \int v\,dt$

$= 3t^2 - \dfrac{t^3}{3} + c$, where c is a constant of integration.

When $t = 0$, $x = 5$

$5 = 3 \times 0^2 - \dfrac{0^3}{3} + c = c \Rightarrow c = 5$

The displacement of the particle from O

after t seconds is $\left(3t^2 - \dfrac{t^3}{3} + 5\right)$ m.

b Using the result in **a**, when $t = 6$

$x = 3 \times 6^2 - \dfrac{6^3}{3} + 5 = 41$

The distance from the starting point is

$(41 - 5)\,\text{m} = 36\,\text{m}$.

> You integrate the velocity to find the displacement. You must remember to add the constant of integration. ← **Pure Year 1, Section 13.1**

> This information enables you to find the value of the constant of integration.
> ← **Pure Year 1, Section 13.3**

> This calculation shows you that, when $t = 6$, the particle is 41 m from O. When the particle started, it was 5 m from O. So the distance from its starting point is $(41 - 5)$ m.

Example 7

A particle travels in a straight line. After t seconds its velocity, $v\,\text{m s}^{-1}$, is given by $v = 5 - 3t^2$, $t \geq 0$. Find the distance travelled by the particle in the third second of its motion.

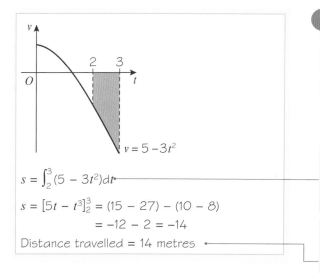

> **Watch out** Before using definite integration to find the distance travelled, check that v doesn't **change sign** in the interval you are considering. A sketch of the velocity–time graph can help.

$s = \int_2^3 (5 - 3t^2)\,dt$

$s = [5t - t^3]_2^3 = (15 - 27) - (10 - 8)$

$\qquad = -12 - 2 = -14$

Distance travelled = 14 metres

> The distance travelled is the area under the velocity–time graph. Use definite integration to find it. ← **Pure Year 1, Section 13.4**

> The velocity is negative between $t = 2$ and $t = 3$ so the **displacement** will be negative. You are asked to find the distance travelled so give the positive numerical value of the displacement.

Exercise 11D

1 A particle is moving in a straight line. Given that $s = 0$ when $t = 0$, find an expression for the displacement of the particle if the velocity is given by:

  a $v = 3t^2 - 1$ **b** $v = 2t^3 - \dfrac{3t^2}{2}$ **c** $v = 2\sqrt{t} + 4t^2$

2 A particle is moving in a straight line. Given that $v = 0$ when $t = 0$, find an expression for the velocity of the particle if the acceleration is given by:

a $a = 8t - 2t^2$

b $a = 6 + \dfrac{t^2}{3}$

3 A particle P is moving on the x-axis. At time t seconds, the velocity of P is $(8 + 2t - 3t^2)\,\text{m s}^{-1}$ in the direction of x increasing. At time $t = 0$, P is at the point where $x = 4$. Find the distance of P from O when $t = 1$.

4 A particle P is moving on the x-axis. At time t seconds, the acceleration of P is $(16 - 2t)\,\text{m s}^{-2}$ in the direction of x increasing. The velocity of P at time t seconds is $v\,\text{m s}^{-1}$. When $t = 0$, $v = 6$ and when $t = 3$, $x = 75$. Find:

a v in terms of t

b the value of x when $t = 0$.

P **5** A particle is moving in a straight line. At time t seconds, its velocity, $v\,\text{m s}^{-1}$, is given by $v = 6t^2 - 51t + 90$. When $t = 0$ the displacement is 0. Find the distance between the two points where P is instantaneously at rest.

P **6** At time t seconds, where $t \geqslant 0$, the velocity $v\,\text{m s}^{-1}$ of a particle moving in a straight line is given by $v = 12 + t - 6t^2$. When $t = 0$, P is at a point O on the line. Find the distance of P from O when $v = 0$.

P **7** A particle P is moving on the x-axis. At time t seconds, the velocity of P is $(4t - t^2)\,\text{m s}^{-1}$ in the direction of x increasing. At time $t = 0$, P is at the origin O. Find:

a the value of x at the instant when $t > 0$ and P is at rest

b the total distance moved by P in the interval $0 \leqslant t \leqslant 5$.

> **Problem-solving**
>
> You will need to consider the motion when v is positive and negative separately.

P **8** A particle P is moving on the x-axis. At time t seconds, the velocity of P is $(6t^2 - 26t + 15)\,\text{m s}^{-1}$ in the direction of x increasing. At time $t = 0$, P is at the origin O. In the subsequent motion P passes through O twice. Find the two non-zero values of t when P passes through O.

P **9** A particle P moves along the x-axis. At time t seconds (where $t \geqslant 0$) the velocity of P is $(3t^2 - 12t + 5)\,\text{m s}^{-1}$ in the direction of x increasing. When $t = 0$, P is at the origin O. Find:

a the values of t when P is again at O

b the distance travelled by P in the interval $2 \leqslant t \leqslant 3$.

P **10** A particle P moves on the x-axis. The acceleration of P at time t seconds, $t \geqslant 0$, is $(4t - 3)\,\text{m s}^{-2}$ in the positive x-direction. When $t = 0$, the velocity of P is $4\,\text{m s}^{-1}$ in the positive x-direction. When $t = T\ (T \neq 0)$, the velocity of P is $4\,\text{m s}^{-1}$ in the positive x-direction. Find the value of T. **(6 marks)**

(E) **11** A particle P travels in a straight line such that its acceleration at time t seconds is $(t - 3)\,\text{m s}^{-2}$. The velocity of P at time t seconds is $v\,\text{m s}^{-1}$. When $t = 0$, $v = 4$. Find:

    a v in terms of t **(4 marks)**

    b the values of t when P is instantaneously at rest   **(3 marks)**

    c the distance between the two points at which P is instantaneously at rest.   **(4 marks)**

(E/P) **12** A particle travels in a straight line such that its acceleration, $a\,\text{m s}^{-2}$, at time t seconds is given by $a = 6t + 2$. When $t = 2$ seconds, the displacement, s, is 10 metres and when $t = 3$ seconds the displacement is 38 metres. Find:

    a the displacement when $t = 4$ seconds   **(6 marks)**

    b the velocity when $t = 4$ seconds.   **(2 marks)**

> **Problem-solving**
>
> You need to use integration to find expressions for the velocity and displacement then substitute in the given values. Use simultaneous equations to find the values of the constants of integration.

> **Challenge**
>
> The motion of a robotic arm moving along a straight track is modelled using the equations:
>
> $$v = \frac{t^2}{2} + 2,\ 0 \leqslant t \leqslant k \quad \text{and} \quad v = 10 + \frac{t}{3} - \frac{t^2}{12},\ k \leqslant t \leqslant 10$$
>
> The diagram shows a sketch of the velocity–time graph of the motion of the arm.
>
>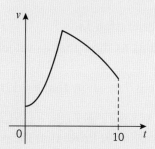
>
> Work out the total distance travelled by the robotic arm.

11.5 Constant acceleration formulae

You can use calculus to derive the formulae for motion with constant acceleration.

> **Example** **8**

A particle moves in a straight line with constant acceleration, $a\,\text{m s}^{-2}$. Given that its initial velocity is $u\,\text{m s}^{-1}$ and its initial displacement is $0\,\text{m}$, prove that:

a its velocity, $v\,\text{m s}^{-1}$ at time $t\,\text{s}$ is given by $v = u + at$

b its displacement, $s\,\text{m}$, at time t is given by $s = ut + \frac{1}{2}at^2$

a $v = \int a \, dt$

$= at + c$

When $t = 0$, $v = u$,

so $u = a \times 0 + c = c$ ——————— Use the initial condition you are given for the velocity to work out the value of c.

So $v = u + at$

b $s = \int v \, dt$

$= \int (u + at) \, dt$ ——————— Use the equation for velocity you have just proved.

$= ut + \frac{1}{2}at^2 + c$

When $t = 0$, $s = 0$ ——————— Use the initial condition you are given for the displacement to work out the value of c.

so $0 = u \times 0 + \frac{1}{2} \times a \times 0^2 + c$

$c = 0$

So $s = ut + \frac{1}{2}at^2$

Watch out The *suvat* equations can only be used when the acceleration is constant.

Exercise 11E

P 1 A particle moves on the x-axis with constant acceleration a m s^{-2}. The particle has initial velocity 0 and initial displacement x m. After time t seconds the particle has velocity v m s^{-1} and displacement s m.

Prove that $s = \frac{1}{2}at^2 + x$.

2 A particle moves in a straight line with constant acceleration 5 m s^{-2}.

  a Given that its initial velocity is 12 m s^{-1}, use calculus to show that its velocity at time t s is given by $v = 12 + 5t$.

  b Given that the initial displacement of the particle is 7 m, show that $s = 12t + 2.5t^2 + 7$.

P 3 A particle moves in a straight line from a point O. At time t seconds, its displacement, s m, from P is given by $s = ut + \frac{1}{2}at^2$ where u and a are constants. Prove that the particle moves with constant acceleration a.

4 Which of these equations for displacement describe constant acceleration? Explain your answers.

  A $s = 2t^2 - t^3$ B $s = 4t + 7$ C $s = \dfrac{t^2}{4}$ D $s = 3t - \dfrac{2}{t^2}$ E $s = 6$

E/P 5 A particle moves in a straight line with constant acceleration. The initial velocity of the particle is 5 m s^{-1} and after 2 seconds it is moving with velocity 13 m s^{-1}.

  a Find the acceleration of the particle. **(3 marks)**

  b Without making use of the kinematics formulae, show that the displacement, s m, of the particle from its starting position is given by:

$$s = pt^2 + qt + r, \; t \geq 0$$

  where p, q and r are constants to be found. **(5 marks)**

Watch out An exam question might specify that you cannot use certain formulae or techniques. In this case you need to use calculus to find the answer to part **b**.

E/P 6 A train travels along a straight track, passing point A at time $t = 0$ and passing point B 40 seconds later. Its distance from A at time t seconds is given by:

$$s = 25t - 0.2t^2, 0 \leqslant t \leqslant 40$$

a Find the distance AB. **(1 mark)**

b Show that the train travels with constant acceleration. **(3 marks)**

A bird passes point B at time $t = 0$ at an initial velocity towards A of $7\,\text{m s}^{-1}$. It flies in a straight line towards point A with constant acceleration $0.6\,\text{m s}^{-2}$.

c Find the distance from A at which the bird is directly above the train. **(6 marks)**

Mixed exercise 11

1 A particle P moves in a horizontal straight line. At time t seconds (where $t \geqslant 0$) the velocity $v\,\text{m s}^{-1}$ of P is given by $v = 15 - 3t$. Find:

a the value of t when P is instantaneously at rest

b the distance travelled by P between the time when $t = 0$ and the time when P is instantaneously at rest.

2 A particle P moves along the x-axis so that, at time t seconds, the displacement of P from O is x metres and the velocity of P is $v\,\text{m s}^{-1}$, where:

$$v = 6t + \tfrac{1}{2}t^3$$

a Find the acceleration of P when $t = 4$.

b Given also that $x = -5$ when $t = 0$, find the distance OP when $t = 4$.

P 3 A particle P is moving along a straight line. At time $t = 0$, the particle is at a point A and is moving with velocity $8\,\text{m s}^{-1}$ towards a point B on the line, where $AB = 30\,\text{m}$. At time t seconds (where $t \geqslant 0$), the acceleration of P is $(2 - 2t)\,\text{m s}^{-2}$ in the direction $\overrightarrow{AB}$.

a Find an expression, in terms of t, for the displacement of P from A at time t seconds.

b Show that P does not reach B.

c Find the value of t when P returns to A, giving your answer to 3 significant figures.

d Find the total distance travelled by P in the interval between the two instants when it passes through A.

E 4 A particle starts from rest at a point O and moves along a straight line OP with an acceleration, a, after t seconds given by $a = (8 - 2t^2)\,\text{m s}^{-2}$.

Find:

a the greatest speed of the particle in the direction OP **(5 marks)**

b the distance covered by the particle in the first two seconds of its motion. **(4 marks)**

E/P 5 A particle P passes through a point O and moves in a straight line. The displacement, s metres, of P from O, t seconds after passing through O is given by:

$$s = -t^3 + 11t^2 - 24t$$

a Find an expression for the velocity, $v\,\text{m s}^{-1}$, of P at time t seconds. **(2 marks)**

b Calculate the values of t at which P is instantaneously at rest. **(3 marks)**

  c Find the value of t at which the acceleration is zero. **(2 marks)**

  d Sketch a velocity–time graph to illustrate the motion of P in the interval $0 \leqslant t \leqslant 6$, showing on your sketch the coordinates of the points at which the graph crosses the axes. **(3 marks)**

  e Calculate the values of t in the interval $0 \leqslant t \leqslant 6$ between which the speed of P is greater than $16\,\text{m s}^{-1}$. **(6 marks)**

(E) **6** A body moves in a straight line. Its velocity, $v\,\text{m s}^{-1}$, at time t seconds is given by $v = 3t^2 - 11t + 10$. Find:

  a the values of t when the body is instantaneously at rest **(3 marks)**

  b the acceleration of the body when $t = 4$ **(3 marks)**

  c the total distance travelled by the body in the interval $0 \leqslant t \leqslant 4$. **(4 marks)**

(E) **7** A particle moves along the positive x-axis. At time $t = 0$ the particle passes through the origin with velocity $6\,\text{m s}^{-1}$. The acceleration, $a\,\text{m s}^{-2}$, of the particle at time t seconds is given by $a = 2t^3 - 8t$ for $t \geqslant 0$. Find:

  a the velocity of the particle at time t seconds **(3 marks)**

  b the displacement of the particle from the origin at time t seconds **(2 marks)**

  c the values of t at which the particle is instantaneously at rest. **(3 marks)**

(E) **8** A remote control drone hovers such that its vertical height, $s\,\text{m}$, above ground level at time t seconds is given by the equation:

$$x = \frac{t^4 - 12t^3 + 28t^2 + 400}{50}, \quad 0 \leqslant t \leqslant 8$$

The diagram shows a sketch of a displacement–time graph of the drone's motion.

Determine the maximum and minimum height of the drone.

(7 marks)

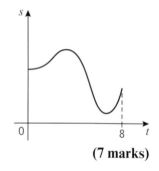

(E/P) **9** A rocket sled is used to test a parachute braking mechanism for a space capsule.

At the moment the parachute is deployed, the sled is 1.5 km from its launch site and is travelling away from it at a speed of $800\,\text{m s}^{-1}$. The sled comes to rest 25 seconds after the parachute is deployed.

The rocket sled is modelled as a particle moving in a straight horizontal line with constant acceleration. At a time t seconds after the parachute is deployed, its distance, $s\,\text{m}$, from the launch site is given by:

$$s = a + bt + ct^2, \, 0 \leqslant t \leqslant 25$$

Find the values of a, b and c in this model. **(6 marks)**

(E) **10** A particle P moves along the x-axis. It passes through the origin O at time $t = 0$ with speed $7\,\text{m s}^{-1}$ in the direction of x increasing.

At time t seconds the acceleration of P in the direction of x increasing is $(20 - 6t)\,\text{m s}^{-2}$.

  a Show that the velocity $v\,\text{m s}^{-1}$ of P at time t seconds is given by:

$$v = 7 + 20t - 3t^2$$

(3 marks)

b Show that $v = 0$ when $t = 7$ and find the greatest speed of P in the interval $0 \leqslant t \leqslant 7$. **(4 marks)**

c Find the distance travelled by P in the interval $0 \leqslant t \leqslant 7$. **(4 marks)**

(E/P) **11** A particle P moves along a straight line. Initially, P is at rest at a point O on the line. At time t seconds (where $t \geqslant 0$) the acceleration of P is proportional to $(7 - t^2)$ and the displacement of P from O is s metres. When $t = 3$, the velocity of P is $6\,\text{m s}^{-1}$.

Show that $s = \frac{1}{24}t^2(42 - t^2)$. **(7 marks)**

(E/P) **12** A mouse leaves its hole and makes a short journey along a straight wall before returning to its hole. The mouse is modelled as a particle moving in a straight line. The distance of the mouse, s m, from its hole at time t minutes is given by:

$$s = t^4 - 10t^3 + 25t^2,\ 0 \leqslant t \leqslant 5$$

a Explain the restriction $0 \leqslant t \leqslant 5$. **(3 marks)**

b Find the greatest distance of the mouse from its hole. **(6 marks)**

(P) **13** At a time t seconds after launch, the space shuttle can be modelled as a particle moving in a straight line with acceleration, $a\,\text{m s}^{-2}$, given by the equation:

$$a = (6.77 \times 10^{-7})t^3 - (3.98 \times 10^{-4})t^2 + 0.105t + 0.859, \quad 124 \leqslant t \leqslant 446$$

a Suggest two reasons why the space shuttle might experience variable acceleration during its launch phase.

Given that the velocity of the space shuttle at time $t = 124$ is $974\,\text{m s}^{-1}$:

b find an expression for the velocity $v\,\text{m s}^{-1}$ of the space shuttle at time t. Give your coefficients to 3 significant figures.

c Hence find the velocity of the space shuttle at time $t = 446$, correct to 3 s.f.

From $t = 446$, the space shuttle maintains a constant acceleration of $28.6\,\text{m s}^{-2}$ until it reaches its escape velocity of $7.85\,\text{km s}^{-1}$. It then cuts its main engines.

d Calculate the time at which the space shuttle cuts its main engines.

Challenge

1 A particle starts at rest and moves in a straight line. At time t seconds after the beginning of its motion, the acceleration of the particle, $a\,\text{m s}^{-2}$, is given by:

$$a = 3t^2 - 18t + 20,\ t \geqslant 0$$

Find the distance travelled by the particle in the first 5 seconds of its motion.

2 A particle travels in a straight line with an acceleration, $a\,\text{m s}^{-2}$, given by $a = 6t + 2$.

The particle travels 50 metres in the fourth second. Find the velocity of the particle when $t = 5$ seconds.

Summary of key points

1 If the displacement, s, is expressed as a function of t, then the velocity, v, can be expressed as

$$v = \frac{ds}{dt}$$

2 If the velocity, v, is expressed as a function of t, then the acceleration, a, can be expressed as

$$a = \frac{dv}{dt} = \frac{d^2s}{dt^2}$$

3

Differentiate | Integrate

$$\text{displacement} \quad = s = \int v\, dt$$

$$\frac{ds}{dt} = \text{velocity} \quad = v = \int a\, dt$$

$$\frac{dv}{dt} = \frac{d^2s}{dt^2} = \text{acceleration} = a$$

Review exercise

E **1**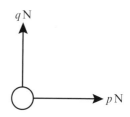

v(m s⁻¹)

The figure shows the velocity–time graph of a cyclist moving on a straight road over a 7 s period. The sections of the graph from $t = 0$ to $t = 3$, and from $t = 3$ to $t = 7$, are straight lines, The section from $t = 3$ to $t = 7$ is parallel to the *t*-axis.

State what can be deduced about the motion of the cyclist from the fact that:

a the graph from $t = 0$ to $t = 3$ is a straight line **(1)**

b the graph from $t = 3$ to $t = 7$ is parallel to the *t*-axis. **(1)**

c Find the distance travelled by the cyclist during this 7 s period. **(4)**

← Section 9.2

E **2** A train stops at two stations 7.5 km apart. Between the stations it takes 75 s to accelerate uniformly to a speed 24 m s⁻¹, then travels at this speed for a time *T* seconds before decelerating uniformly for the final 0.6 km.

a Draw a velocity–time graph to illustrate this journey. **(3)**

Hence, or otherwise, find:

b the deceleration of the train during the final 0.6 km **(3)**

c the value of *T* **(5)**

d the total time for the journey. **(4)**

← Sections 9.2, 9.3

E/P **3** An electric train starts from rest at a station *A* and moves along a straight level track. The train accelerates uniformly at 0.4 m s⁻² to a speed of 16 m s⁻¹. The speed is then maintained for a distance of 2000 m. Finally the train retards uniformly for 20 s before coming to rest at a station *B*. For this journey from *A* to *B*,

a find the total time taken **(5)**

b find the distance from *A* to *B* **(5)**

c sketch the displacement–time graph, showing clearly the shape of the graph for each stage of the journey. **(3)**

← Sections 9.1, 9.3

E **4** A small ball is projected vertically upwards from a point *A*. The greatest height reached by the ball is 40 m above *A*. Calculate:

a the speed of projection **(3)**

b the time between the instant that the ball is projected and the instant it returns to *A*. **(3)**

← Sections 9.4, 9.5

E/P **5** A ball is projected vertically upwards and takes 3 seconds to reach its highest point. At time *t* seconds, the ball is 39.2 m above its point of projection. Find the possible values of *t*. **(5)**

← Sections 9.4, 9.5

E/P **6** A light object is acted upon by a horizontal force of *p* N and a vertical force of *q* N as shown in the diagram.

q N

p N

The resultant of the two forces has a magnitude of $\sqrt{40}\,\text{N}$ which acts in the direction of $30°$ to the horizontal. Calculate the value of p and the value of q.

← **Sections 8.4, 10.1, 10.2**

(E) **7** A car of mass $750\,\text{kg}$, moving along a level straight road, has its speed reduced from $25\,\text{m s}^{-1}$ to $15\,\text{m s}^{-1}$ by brakes which produce a constant retarding force of $2250\,\text{N}$. Calculate the distance travelled by the car as its speed is reduced from $25\,\text{m s}^{-2}$ to $15\,\text{m s}^{-1}$. **(5)**

← **Sections 9.3, 10.3**

(E) **8** An engine of mass 25 tonnes pulls a truck of mass 10 tonnes along a railway line. The resistances to the motion of the engine and the truck are modelled as constant and of magnitude $50\,\text{N}$ per tonne. When the train is travelling horizontally, the tractive force exerted by the engine is $26\,\text{kN}$. Modelling the engine and the truck as particles and the coupling between the engine and the truck as a light horizontal rod calculate:

a the acceleration of the engine and the truck **(4)**

b the tension in the coupling. **(3)**

c State how in your calculations you have used the information that
  i the engine and the truck are particles
  ii the coupling is a light horizontal rod. **(2)**

← **Sections 8.1, 8.2, 10.3, 10.5**

(E/P) **9** A ball is projected vertically upwards with a speed $u\,\text{m s}^{-1}$ from a point A, which is $1.5\,\text{m}$ above the ground. The ball moves freely under gravity until it reaches the ground. The greatest height attained by the ball is $25.6\,\text{m}$ above A.

a Show that $u = 22.4$. **(3)**

The ball reaches the ground T seconds after it has been projected from A.

b Find, to three significant figures, the value of T. **(3)**

The ground is soft and the ball sinks $2.5\,\text{cm}$ into the ground before coming to rest. The mass of the ball is $0.6\,\text{kg}$. The ground is assumed to exert a constant resistive force of magnitude F newtons.

c Find, to three significant figures, the value of F. **(4)**

d Sketch a velocity–time graph for the entire motion of the ball, showing the values of t at any points where the graph intercepts the horizontal axis. **(4)**

e State one physical factor which could be taken into account to make the model used in this question more realistic. **(1)**

← **Sections 8.1, 8.2, 9.5, 10.3, 10.4**

(E) **10** A particle A, of mass $0.8\,\text{kg}$, resting on a smooth horizontal table, is connected to a particle B, of mass $0.6\,\text{kg}$, which is $1\,\text{m}$ from the ground, by a light inextensible string passing over a small smooth pulley at the edge of the table. The particle A is more than $1\,\text{m}$ from the edge of the table. The system is released from rest with the horizontal part of the string perpendicular to the edge of the table, the hanging parts vertical and the string taut.

Calculate:

a the acceleration of A **(5)**

b the tension in the string **(1)**

c the speed of B when it hits the ground **(3)**

d the time taken for B to reach the ground. **(3)**

e The string in this question is described as being 'light'.
  i Write down what you understand by this description.
  ii State how you have used the fact that the string is light in your answer to parts **a** and **b**. **(2)**

← **Sections 8.1, 8.2, 9.5, 10.4, 10.6**

(E/P) **11** Two particles P and Q have mass $0.6\,\text{kg}$ and $0.2\,\text{kg}$ respectively. The particles are attached to the ends of a light inextensible string. The string passes over a small smooth pulley which is fixed above a horizontal floor. Both particles are held, with the string taut, at a height of 1 m above the floor. The particles are released from rest and in the subsequent motion Q does not reach the pulley.

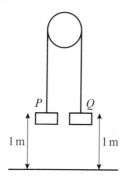

a Find the tension in the string immediately after the particles are released. **(6)**

b Find the acceleration of P immediately after the particles are released. **(2)**

When the particles have been moving for 0.4 s, the string breaks.

c Find the further time that elapses until P hits the floor. **(9)**

d State how in your calculations you have used the information that the string is inextensible. **(1)**

← Sections 9.5, 10.4, 10.6

E **12** A trailer of mass 600 kg is attached to a car of mass 900 kg by means of a light inextensible tow-bar. The car tows the trailer along a horizontal road. The resistances to motion of the car and trailer are 300 N and 150 N respectively.

a Given that the acceleration of the car and trailer is 0.4 m s⁻², calculate:

i the tractive force exerted by the engine of the car

ii the tension in the tow-bar. **(6)**

b Given that the magnitude of the force in the tow-bar must not exceed 1650 N, calcluate the greatest possible deceleration of the car. **(3)**

← Sections 10.1, 10.3, 10.5

E **13** A boy sits on a box in a lift. The mass of the boy is 45 kg, the mass of the box is 20 kg and the mass of the lift is 1050 kg. The lift is being raised vertically by a vertical cable which is attached to the top of the lift. The lift is moving upwards and has constant deceleration of 2 m s⁻². By modelling the cable as being light and inextensible, find:

a the tension in the cable **(3)**

b the magnitude of the force exerted on the box by the boy **(3)**

c the magnitude of the force exerted on the box by the lift. **(3)**

← Sections 8.1, 8.2, 10.1, 10.4, 10.5

E/P **14** Two forces $F_1 = (2\mathbf{i} + 3\mathbf{j})$ N and $F_2 = (\lambda\mathbf{i} + \mu\mathbf{j})$ N, where λ and μ are scalars, act on a particle. The resultant of the two forces is $\mathbf{R}$, where $\mathbf{R}$ is parallel to the vector $\mathbf{i} + 2\mathbf{j}$.

a Find, to the nearest degree, the acute angle between the line of action of $\mathbf{R}$ and the vector $\mathbf{i}$. **(2)**

b Show that $2\lambda - \mu + 1 = 0$. **(5)**

Given that the direction of F_2 is parallel to $\mathbf{j}$,

c find, to three significant figures the magnitude of $\mathbf{R}$. **(4)**

← Sections 8.4, 10.2

E/P **15** A force $\mathbf{R}$ acts on a particle, where $\mathbf{R} = (7\mathbf{i} + 16\mathbf{j})$ N.

Calculate:

a the magnitude of $\mathbf{R}$, giving your answers to one decimal place **(2)**

b the angle between the line of action of $\mathbf{R}$ and $\mathbf{i}$, giving your answer to the nearest degree. **(2)**

The force **R** is the resultant of two forces **P** and **Q**. The line of action of **P** is parallel to the vector $(\mathbf{i} + 4\mathbf{j})$ and the line of action of **Q** is parallel to the vector $(\mathbf{i} + \mathbf{j})$.

  c Determine the forces **P** and **Q** expressing each in terms of **i** and **j**. **(6)**

        ← **Sections 8.4, 10.2**

(E/P) 16 A particle P moves on the x-axis. At time t seconds, its acceleration is $(5 - 2t)\,\mathrm{m\,s^{-2}}$, measured in the direction of x increasing. When $t = 0$, its velocity is $6\,\mathrm{m\,s^{-1}}$ measured in the direction of x increasing. Find the time when P is instantaneously at rest in the subsequent motion. **(5)**

        ← **Sections 11.1, 11.4**

(E/P) 17 At time $t = 0$ a particle P leaves the origin O and moves along the x-axis. At time t seconds the velocity of P is $v\,\mathrm{m\,s^{-1}}$, where $v = 6t - 2t^2$. Find:

  a the maximum value of v **(4)**

  b the time taken for P to return to O. **(5)**

        ← **Sections 11.1, 11.2, 11.3, 11.4**

(E/P) 18 A particle P moves on the positive x-axis. The velocity of P at time t seconds is $(3t^2 - 8t + 5)\,\mathrm{m\,s^{-1}}$. When $t = 0$, P is $12\,\mathrm{m}$ from the origin O. Find:

  a the values of t when P is instantaneously at rest **(3)**

  b the acceleration of P when $t = 4$ **(3)**

  c the total distance travelled by P in the third second. **(4)**

        ← **Sections 11.1, 11.2, 11.3, 11.4**

(E) 19 A particle moves in a straight line and at time t seconds has velocity $v\,\mathrm{m\,s^{-1}}$, where

$$v = 6t - 2t^{\frac{3}{2}}, \; t \geqslant 0$$

  a Find an expression for the acceleration of the particle at time t. **(2)**

When $t = 0$, the particle is at the origin.

  b Find an expression for the displacement of the particle from the origin at time t. **(4)**

        ← **Sections 11.1, 11.2, 11.3, 11.4**

Challenge

1 A tram starts from rest at station A and accelerates uniformly for t_1 seconds covering a distance of 1750 m. It then travels at a constant speed $v\,\mathrm{m\,s^{-1}}$ for t_2 seconds covering a distance of 17 500 m. The tram then decelerates for t_3 seconds and comes to rest at station B. Given that the total time for the journey is 7 minutes and $3t_1 = 4t_3$, find t_1, t_2 and t_3 and the distance between station A and station B.

        ← **Section 9.3**

2 [In this question use $g = 10\,\mathrm{m\,s^{-2}}$]

One end of a light inextensible string is attached to a block A of mass 5 kg. The block A is held at rest on a rough horizontal table. The motion of the block is subject to a resistance of 2 N. The string lies parallel to the table and passes over a smooth light pulley which is fixed at the top of the table. The other end of the string is attached to a light scale pan which carries two blocks B and C, with block B on top of block C as shown. The mass of block B is 5 kg and the mass of block C is 10 kg.

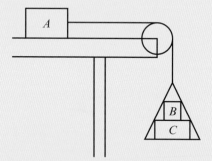

The scale pan hangs at rest and the system is released from rest. By modelling the blocks as particles, ignoring air resistance and assuming the motion is uninterrupted, find:

  a the acceleration of the scale pan

  b the tension in the string

  c the magnitude of the force exerted on block B by block C

  d the magnitude of the force exerted on the pulley by the string.

  e State how in your calculations you have used the information that the string is inextensible.

        ← **Sections 10.5, 10.6**

Exam-style practice

Mathematics
AS Level
Paper 2: Statistics and Mechanics

Time: 1 hour 15 minutes
You must have: Mathematical Formulae and Statistical Tables, Calculator

SECTION A: STATISTICS

1   The Venn diagram shows the probabilities that a randomly chosen member of a group of monkeys likes bananas (B) or mangoes (M).

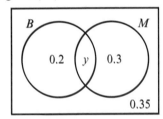

   a Find the value of y. **(1)**

   b Determine whether the events 'likes bananas' and 'likes mangoes' are independent. **(2)**

2   Clare is investigating the daily mean temperature in the UK in September 2015. She takes a sample of the first 10 days from September 2015 for Camborne from the large data set. The results are shown below:

   14.3   12.8   13.0   13.0   14.3   12.6   13.5   13.7   15.9   17.0

   a State, with a reason, whether t is a discrete or continuous variable. **(1)**

   Given that $\Sigma t = 140.1$ and $\Sigma t^2 = 1981.33$,

   b find the mean and standard deviation of the temperatures. **(3)**

   The mean temperature on 11 September is recorded as 15.8 °C.

   c State what effect adding this value to the data set would have on the mean temperature. **(1)**

   d Suggest how Clare could make better use of the large data set for her study. **(2)**

3   A biased dice has a probability distribution as shown in the table below:

x	1	2	3	4	5	6
$P(X = x)$	0.1	0.2	0.15	p	0.1	0.25

   a Find the value of p. **(1)**

   b Find $P(2 \leqslant X \leqslant 5)$. **(1)**

   c The dice is rolled 10 times Find the probability that it lands on an odd number:

      i exactly twice **(2)**

      ii more than 6 times. **(2)**

4 A factory makes plates using a production line process. On average, 3 out of every 10 plates have flaws. A new production process is introduced designed to make the average number of flaws less. A new sample of 20 plates is taken.

  a Describe the test statistic and state suitable null and alternative hypotheses. **(2)**

  b Using a 5% level of significance, find the critical region for a test to check the belief that the process has improved. **(3)**

  c State the actual significance level. **(1)**

  In the new sample, only 1 plate has flaws.

  d Conclude whether there is evidence that the process has improved. **(1)**

5 A scientist measures the amount of energy released by a chemical reaction, e Joules, against the temperature, $h\,°C$.

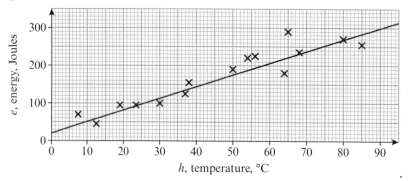

  She found the equation of the regression line of e on h to be $e = 20 + 3.1h$.

  a Give an interpretation of the value 3.1 in this model. **(1)**

  b State, with a reason, whether it is sensible to estimate e when $h = 200\,°C$. **(1)**

  c State, with a reason, whether it is sensible to measure h when $e = 150$ Joules. **(1)**

6 A conservationist is collecting data on the heights of giraffe. She displays the data in a histogram as shown.

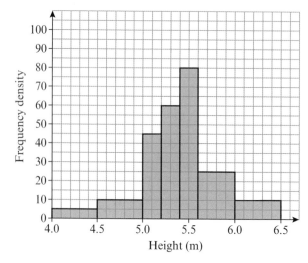

  One giraffe is chosen at random. Estimate the probability that it is between 4.6 and 6.1 metres tall. **(4)**

SECTION B: MECHANICS

7   A car is towing a trailer along a straight horizontal road by means of a horizontal tow-rope. The mass of the car is 1500 kg. The mass of the trailer is 700 kg. The car and the trailer are modelled as particles and the tow-rope as a light inextensible string. The resistances to motion of the car and the trailer are assumed to be constant and of magnitude 660 N and 320 N respectively. The driving force on the car, due to its engine, is 2630 N.

Find:

a the acceleration of the car   **(3)**

b the tension in the tow-rope.   **(3)**

c State how you have used the modelling assumption that the tow-rope is inextensible.   **(1)**

8   A particle P of mass 3 kg is moving under the action of forces
$\mathbf{F}_1 = 3\mathbf{i} - 6\mathbf{j}$ N, $\mathbf{F}_2 = 4\mathbf{i} + 5\mathbf{j}$ N and $\mathbf{F}_3 = 2\mathbf{i} - 2\mathbf{j}$ N.

Find:

a the acceleration of P in the form $p\mathbf{i} + q\mathbf{j}$ **(3)**

b angle the acceleration makes with $\mathbf{i}$ **(2)**

c the magnitude of the acceleration.   **(2)**

9   A small ball is projected vertically upwards from a point P with speed u m s^{-1}. After projection the ball moves freely under gravity until it returns to P. The time between the instant that the ball is projected and the instant that it returns to P is 5 seconds.

The ball is modelled as a particle moving freely under gravity.

Find:

a the value of u **(3)**

b the greatest height above P reached by the ball.   **(2)**

At time t seconds, the ball is 15 m above P.

c Find the possible values of t.   **(4)**

10   A particle, P, moves in a straight line through a fixed point O. The velocity of the particle, v m s^{-1} at a time t seconds after passing through O is given by

$v = 3 + 9t^2 - 4t^3$, $0 \leqslant t \leqslant 2$.

The diagram shows a velocity–time graph of the motion of P.

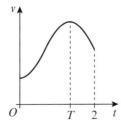

Find the distance of P from O at time T seconds, when the particle is moving with maximum velocity.   **(7)**

Binomial cumulative distribution function

The tabulated value is P($X \leq x$), where X has a binomial distribution with index n and parameter p.

$p =$	0.05	0.10	0.15	0.20	0.25	0.30	0.35	0.40	0.45	0.50
$n = 5, x = 0$	0.7738	0.5905	0.4437	0.3277	0.2373	0.1681	0.1160	0.0778	0.0503	0.0312
1	0.9774	0.9185	0.8352	0.7373	0.6328	0.5282	0.4284	0.3370	0.2562	0.1875
2	0.9988	0.9914	0.9734	0.9421	0.8965	0.8369	0.7648	0.6826	0.5931	0.5000
3	1.0000	0.9995	0.9978	0.9933	0.9844	0.9692	0.9460	0.9130	0.8688	0.8125
4	1.0000	1.0000	0.9999	0.9997	0.9990	0.9976	0.9947	0.9898	0.9815	0.9688
$n = 6, x = 0$	0.7351	0.5314	0.3771	0.2621	0.1780	0.1176	0.0754	0.0467	0.0277	0.0156
1	0.9672	0.8857	0.7765	0.6554	0.5339	0.4202	0.3191	0.2333	0.1636	0.1094
2	0.9978	0.9842	0.9527	0.9011	0.8306	0.7443	0.6471	0.5443	0.4415	0.3438
3	0.9999	0.9987	0.9941	0.9830	0.9624	0.9295	0.8826	0.8208	0.7447	0.6563
4	1.0000	0.9999	0.9996	0.9984	0.9954	0.9891	0.9777	0.9590	0.9308	0.8906
5	1.0000	1.0000	1.0000	0.9999	0.9998	0.9993	0.9982	0.9959	0.9917	0.9844
$n = 7, x = 0$	0.6983	0.4783	0.3206	0.2097	0.1335	0.0824	0.0490	0.0280	0.0152	0.0078
1	0.9556	0.8503	0.7166	0.5767	0.4449	0.3294	0.2338	0.1586	0.1024	0.0625
2	0.9962	0.9743	0.9262	0.8520	0.7564	0.6471	0.5323	0.4199	0.3164	0.2266
3	0.9998	0.9973	0.9879	0.9667	0.9294	0.8740	0.8002	0.7102	0.6083	0.5000
4	1.0000	0.9998	0.9988	0.9953	0.9871	0.9712	0.9444	0.9037	0.8471	0.7734
5	1.0000	1.0000	0.9999	0.9996	0.9987	0.9962	0.9910	0.9812	0.9643	0.9375
6	1.0000	1.0000	1.0000	1.0000	0.9999	0.9998	0.9994	0.9984	0.9963	0.9922
$n = 8, x = 0$	0.6634	0.4305	0.2725	0.1678	0.1001	0.0576	0.0319	0.0168	0.0084	0.0039
1	0.9428	0.8131	0.6572	0.5033	0.3671	0.2553	0.1691	0.1064	0.0632	0.0352
2	0.9942	0.9619	0.8948	0.7969	0.6785	0.5518	0.4278	0.3154	0.2201	0.1445
3	0.9996	0.9950	0.9786	0.9437	0.8862	0.8059	0.7064	0.5941	0.4770	0.3633
4	1.0000	0.9996	0.9971	0.9896	0.9727	0.9420	0.8939	0.8263	0.7396	0.6367
5	1.0000	1.0000	0.9998	0.9988	0.9958	0.9887	0.9747	0.9502	0.9115	0.8555
6	1.0000	1.0000	1.0000	0.9999	0.9996	0.9987	0.9964	0.9915	0.9819	0.9648
7	1.0000	1.0000	1.0000	1.0000	1.0000	0.9999	0.9998	0.9993	0.9983	0.9961
$n = 9, x = 0$	0.6302	0.3874	0.2316	0.1342	0.0751	0.0404	0.0207	0.0101	0.0046	0.0020
1	0.9288	0.7748	0.5995	0.4362	0.3003	0.1960	0.1211	0.0705	0.0385	0.0195
2	0.9916	0.9470	0.8591	0.7382	0.6007	0.4628	0.3373	0.2318	0.1495	0.0898
3	0.9994	0.9917	0.9661	0.9144	0.8343	0.7297	0.6089	0.4826	0.3614	0.2539
4	1.0000	0.9991	0.9944	0.9804	0.9511	0.9012	0.8283	0.7334	0.6214	0.5000
5	1.0000	0.9999	0.9994	0.9969	0.9900	0.9747	0.9464	0.9006	0.8342	0.7461
6	1.0000	1.0000	1.0000	0.9997	0.9987	0.9957	0.9888	0.9750	0.9502	0.9102
7	1.0000	1.0000	1.0000	1.0000	0.9999	0.9996	0.9986	0.9962	0.9909	0.9805
8	1.0000	1.0000	1.0000	1.0000	1.0000	1.0000	0.9999	0.9997	0.9992	0.9980
$n = 10, x = 0$	0.5987	0.3487	0.1969	0.1074	0.0563	0.0282	0.0135	0.0060	0.0025	0.0010
1	0.9139	0.7361	0.5443	0.3758	0.2440	0.1493	0.0860	0.0464	0.0233	0.0107
2	0.9885	0.9298	0.8202	0.6778	0.5256	0.3828	0.2616	0.1673	0.0996	0.0547
3	0.9990	0.9872	0.9500	0.8791	0.7759	0.6496	0.5138	0.3823	0.2660	0.1719
4	0.9999	0.9984	0.9901	0.9672	0.9219	0.8497	0.7515	0.6331	0.5044	0.3770
5	1.0000	0.9999	0.9986	0.9936	0.9803	0.9527	0.9051	0.8338	0.7384	0.6230
6	1.0000	1.0000	0.9999	0.9991	0.9965	0.9894	0.9740	0.9452	0.8980	0.8281
7	1.0000	1.0000	1.0000	0.9999	0.9996	0.9984	0.9952	0.9877	0.9726	0.9453
8	1.0000	1.0000	1.0000	1.0000	1.0000	0.9999	0.9995	0.9983	0.9955	0.9893
9	1.0000	1.0000	1.0000	1.0000	1.0000	1.0000	1.0000	0.9999	0.9997	0.9990

$p =$	0.05	0.10	0.15	0.20	0.25	0.30	0.35	0.40	0.45	0.50
$n = 12, x = 0$	0.5404	0.2824	0.1422	0.0687	0.0317	0.0138	0.0057	0.0022	0.0008	0.0002
1	0.8816	0.6590	0.4435	0.2749	0.1584	0.0850	0.0424	0.0196	0.0083	0.0032
2	0.9804	0.8891	0.7358	0.5583	0.3907	0.2528	0.1513	0.0834	0.0421	0.0193
3	0.9978	0.9744	0.9078	0.7946	0.6488	0.4925	0.3467	0.2253	0.1345	0.0730
4	0.9998	0.9957	0.9761	0.9274	0.8424	0.7237	0.5833	0.4382	0.3044	0.1938
5	1.0000	0.9995	0.9954	0.9806	0.9456	0.8822	0.7873	0.6652	0.5269	0.3872
6	1.0000	0.9999	0.9993	0.9961	0.9857	0.9614	0.9154	0.8418	0.7393	0.6128
7	1.0000	1.0000	0.9999	0.9994	0.9972	0.9905	0.9745	0.9427	0.8883	0.8062
8	1.0000	1.0000	1.0000	0.9999	0.9996	0.9983	0.9944	0.9847	0.9644	0.9270
9	1.0000	1.0000	1.0000	1.0000	1.0000	0.9998	0.9992	0.9972	0.9921	0.9807
10	1.0000	1.0000	1.0000	1.0000	1.0000	1.0000	0.9999	0.9997	0.9989	0.9968
11	1.0000	1.0000	1.0000	1.0000	1.0000	1.0000	1.0000	1.0000	0.9999	0.9998
$n = 15, x = 0$	0.4633	0.2059	0.0874	0.0352	0.0134	0.0047	0.0016	0.0005	0.0001	0.0000
1	0.8290	0.5490	0.3186	0.1671	0.0802	0.0353	0.0142	0.0052	0.0017	0.0005
2	0.9638	0.8159	0.6042	0.3980	0.2361	0.1268	0.0617	0.0271	0.0107	0.0037
3	0.9945	0.9444	0.8227	0.6482	0.4613	0.2969	0.1727	0.0905	0.0424	0.0176
4	0.9994	0.9873	0.9383	0.8358	0.6865	0.5155	0.3519	0.2173	0.1204	0.0592
5	0.9999	0.9978	0.9832	0.9389	0.8516	0.7216	0.5643	0.4032	0.2608	0.1509
6	1.0000	0.9997	0.9964	0.9819	0.9434	0.8689	0.7548	0.6098	0.4522	0.3036
7	1.0000	1.0000	0.9994	0.9958	0.9827	0.9500	0.8868	0.7869	0.6535	0.5000
8	1.0000	1.0000	0.9999	0.9992	0.9958	0.9848	0.9578	0.9050	0.8182	0.6964
9	1.0000	1.0000	1.0000	0.9999	0.9992	0.9963	0.9876	0.9662	0.9231	0.8491
10	1.0000	1.0000	1.0000	1.0000	0.9999	0.9993	0.9972	0.9907	0.9745	0.9408
11	1.0000	1.0000	1.0000	1.0000	1.0000	0.9999	0.9995	0.9981	0.9937	0.9824
12	1.0000	1.0000	1.0000	1.0000	1.0000	1.0000	0.9999	0.9997	0.9989	0.9963
13	1.0000	1.0000	1.0000	1.0000	1.0000	1.0000	1.0000	1.0000	0.9999	0.9995
14	1.0000	1.0000	1.0000	1.0000	1.0000	1.0000	1.0000	1.0000	1.0000	1.0000
$n = 20, x = 0$	0.3585	0.1216	0.0388	0.0115	0.0032	0.0008	0.0002	0.0000	0.0000	0.0000
1	0.7358	0.3917	0.1756	0.0692	0.0243	0.0076	0.0021	0.0005	0.0001	0.0000
2	0.9245	0.6769	0.4049	0.2061	0.0913	0.0355	0.0121	0.0036	0.0009	0.0002
3	0.9841	0.8670	0.6477	0.4114	0.2252	0.1071	0.0444	0.0160	0.0049	0.0013
4	0.9974	0.9568	0.8298	0.6296	0.4148	0.2375	0.1182	0.0510	0.0189	0.0059
5	0.9997	0.9887	0.9327	0.8042	0.6172	0.4164	0.2454	0.1256	0.0553	0.0207
6	1.0000	0.9976	0.9781	0.9133	0.7858	0.6080	0.4166	0.2500	0.1299	0.0577
7	1.0000	0.9996	0.9941	0.9679	0.8982	0.7723	0.6010	0.4159	0.2520	0.1316
8	1.0000	0.9999	0.9987	0.9900	0.9591	0.8867	0.7624	0.5956	0.4143	0.2517
9	1.0000	1.0000	0.9998	0.9974	0.9861	0.9520	0.8782	0.7553	0.5914	0.4119
10	1.0000	1.0000	1.0000	0.9994	0.9961	0.9829	0.9468	0.8725	0.7507	0.5881
11	1.0000	1.0000	1.0000	0.9999	0.9991	0.9949	0.9804	0.9435	0.8692	0.7483
12	1.0000	1.0000	1.0000	1.0000	0.9998	0.9987	0.9940	0.9790	0.9420	0.8684
13	1.0000	1.0000	1.0000	1.0000	1.0000	0.9997	0.9985	0.9935	0.9786	0.9423
14	1.0000	1.0000	1.0000	1.0000	1.0000	1.0000	0.9997	0.9984	0.9936	0.9793
15	1.0000	1.0000	1.0000	1.0000	1.0000	1.0000	1.0000	0.9997	0.9985	0.9941
16	1.0000	1.0000	1.0000	1.0000	1.0000	1.0000	1.0000	1.0000	0.9997	0.9987
17	1.0000	1.0000	1.0000	1.0000	1.0000	1.0000	1.0000	1.0000	1.0000	0.9998
18	1.0000	1.0000	1.0000	1.0000	1.0000	1.0000	1.0000	1.0000	1.0000	1.0000

$p =$	0.05	0.10	0.15	0.20	0.25	0.30	0.35	0.40	0.45	0.50
$n = 25, x = 0$	0.2774	0.0718	0.0172	0.0038	0.0008	0.0001	0.0000	0.0000	0.0000	0.0000
1	0.6424	0.2712	0.0931	0.0274	0.0070	0.0016	0.0003	0.0001	0.0000	0.0000
2	0.8729	0.5371	0.2537	0.0982	0.0321	0.0090	0.0021	0.0004	0.0001	0.0000
3	0.9659	0.7636	0.4711	0.2340	0.0962	0.0332	0.0097	0.0024	0.0005	0.0001
4	0.9928	0.9020	0.6821	0.4207	0.2137	0.0905	0.0320	0.0095	0.0023	0.0005
5	0.9988	0.9666	0.8385	0.6167	0.3783	0.1935	0.0826	0.0294	0.0086	0.0020
6	0.9998	0.9905	0.9305	0.7800	0.5611	0.3407	0.1734	0.0736	0.0258	0.0073
7	1.0000	0.9977	0.9745	0.8909	0.7265	0.5118	0.3061	0.1536	0.0639	0.0216
8	1.0000	0.9995	0.9920	0.9532	0.8506	0.6769	0.4668	0.2735	0.1340	0.0539
9	1.0000	0.9999	0.9979	0.9827	0.9287	0.8106	0.6303	0.4246	0.2424	0.1148
10	1.0000	1.0000	0.9995	0.9944	0.9703	0.9022	0.7712	0.5858	0.3843	0.2122
11	1.0000	1.0000	0.9999	0.9985	0.9893	0.9558	0.8746	0.7323	0.5426	0.3450
12	1.0000	1.0000	1.0000	0.9996	0.9966	0.9825	0.9396	0.8462	0.6937	0.5000
13	1.0000	1.0000	1.0000	0.9999	0.9991	0.9940	0.9745	0.9222	0.8173	0.6550
14	1.0000	1.0000	1.0000	1.0000	0.9998	0.9982	0.9907	0.9656	0.9040	0.7878
15	1.0000	1.0000	1.0000	1.0000	1.0000	0.9995	0.9971	0.9868	0.9560	0.8852
16	1.0000	1.0000	1.0000	1.0000	1.0000	0.9999	0.9992	0.9957	0.9826	0.9461
17	1.0000	1.0000	1.0000	1.0000	1.0000	1.0000	0.9998	0.9988	0.9942	0.9784
18	1.0000	1.0000	1.0000	1.0000	1.0000	1.0000	1.0000	0.9997	0.9984	0.9927
19	1.0000	1.0000	1.0000	1.0000	1.0000	1.0000	1.0000	0.9999	0.9996	0.9980
20	1.0000	1.0000	1.0000	1.0000	1.0000	1.0000	1.0000	1.0000	0.9999	0.9995
21	1.0000	1.0000	1.0000	1.0000	1.0000	1.0000	1.0000	1.0000	1.0000	0.9999
22	1.0000	1.0000	1.0000	1.0000	1.0000	1.0000	1.0000	1.0000	1.0000	1.0000
$n = 30, x = 0$	0.2146	0.0424	0.0076	0.0012	0.0002	0.0000	0.0000	0.0000	0.0000	0.0000
1	0.5535	0.1837	0.0480	0.0105	0.0020	0.0003	0.0000	0.0000	0.0000	0.0000
2	0.8122	0.4114	0.1514	0.0442	0.0106	0.0021	0.0003	0.0000	0.0000	0.0000
3	0.9392	0.6474	0.3217	0.1227	0.0374	0.0093	0.0019	0.0003	0.0000	0.0000
4	0.9844	0.8245	0.5245	0.2552	0.0979	0.0302	0.0075	0.0015	0.0002	0.0000
5	0.9967	0.9268	0.7106	0.4275	0.2026	0.0766	0.0233	0.0057	0.0011	0.0002
6	0.9994	0.9742	0.8474	0.6070	0.3481	0.1595	0.0586	0.0172	0.0040	0.0007
7	0.9999	0.9922	0.9302	0.7608	0.5143	0.2814	0.1238	0.0435	0.0121	0.0026
8	1.0000	0.9980	0.9722	0.8713	0.6736	0.4315	0.2247	0.0940	0.0312	0.0081
9	1.0000	0.9995	0.9903	0.9389	0.8034	0.5888	0.3575	0.1763	0.0694	0.0214
10	1.0000	0.9999	0.9971	0.9744	0.8943	0.7304	0.5078	0.2915	0.1350	0.0494
11	1.0000	1.0000	0.9992	0.9905	0.9493	0.8407	0.6548	0.4311	0.2327	0.1002
12	1.0000	1.0000	0.9998	0.9969	0.9784	0.9155	0.7802	0.5785	0.3592	0.1808
13	1.0000	1.0000	1.0000	0.9991	0.9918	0.9599	0.8737	0.7145	0.5025	0.2923
14	1.0000	1.0000	1.0000	0.9998	0.9973	0.9831	0.9348	0.8246	0.6448	0.4278
15	1.0000	1.0000	1.0000	0.9999	0.9992	0.9936	0.9699	0.9029	0.7691	0.5722
16	1.0000	1.0000	1.0000	1.0000	0.9998	0.9979	0.9876	0.9519	0.8644	0.7077
17	1.0000	1.0000	1.0000	1.0000	0.9999	0.9994	0.9955	0.9788	0.9286	0.8192
18	1.0000	1.0000	1.0000	1.0000	1.0000	0.9998	0.9986	0.9917	0.9666	0.8998
19	1.0000	1.0000	1.0000	1.0000	1.0000	1.0000	0.9996	0.9971	0.9862	0.9506
20	1.0000	1.0000	1.0000	1.0000	1.0000	1.0000	0.9999	0.9991	0.9950	0.9786
21	1.0000	1.0000	1.0000	1.0000	1.0000	1.0000	1.0000	0.9998	0.9984	0.9919
22	1.0000	1.0000	1.0000	1.0000	1.0000	1.0000	1.0000	1.0000	0.9996	0.9974
23	1.0000	1.0000	1.0000	1.0000	1.0000	1.0000	1.0000	1.0000	0.9999	0.9993
24	1.0000	1.0000	1.0000	1.0000	1.0000	1.0000	1.0000	1.0000	1.0000	0.9998
25	1.0000	1.0000	1.0000	1.0000	1.0000	1.0000	1.0000	1.0000	1.0000	1.0000

$p =$	0.05	0.10	0.15	0.20	0.25	0.30	0.35	0.40	0.45	0.50
$n = 40, x = 0$	0.1285	0.0148	0.0015	0.0001	0.0000	0.0000	0.0000	0.0000	0.0000	0.0000
1	0.3991	0.0805	0.0121	0.0015	0.0001	0.0000	0.0000	0.0000	0.0000	0.0000
2	0.6767	0.2228	0.0486	0.0079	0.0010	0.0001	0.0000	0.0000	0.0000	0.0000
3	0.8619	0.4231	0.1302	0.0285	0.0047	0.0006	0.0001	0.0000	0.0000	0.0000
4	0.9520	0.6290	0.2633	0.0759	0.0160	0.0026	0.0003	0.0000	0.0000	0.0000
5	0.9861	0.7937	0.4325	0.1613	0.0433	0.0086	0.0013	0.0001	0.0000	0.0000
6	0.9966	0.9005	0.6067	0.2859	0.0962	0.0238	0.0044	0.0006	0.0001	0.0000
7	0.9993	0.9581	0.7559	0.4371	0.1820	0.0553	0.0124	0.0021	0.0002	0.0000
8	0.9999	0.9845	0.8646	0.5931	0.2998	0.1110	0.0303	0.0061	0.0009	0.0001
9	1.0000	0.9949	0.9328	0.7318	0.4395	0.1959	0.0644	0.0156	0.0027	0.0003
10	1.0000	0.9985	0.9701	0.8392	0.5839	0.3087	0.1215	0.0352	0.0074	0.0011
11	1.0000	0.9996	0.9880	0.9125	0.7151	0.4406	0.2053	0.0709	0.0179	0.0032
12	1.0000	0.9999	0.9957	0.9568	0.8209	0.5772	0.3143	0.1285	0.0386	0.0083
13	1.0000	1.0000	0.9986	0.9806	0.8968	0.7032	0.4408	0.2112	0.0751	0.0192
14	1.0000	1.0000	0.9996	0.9921	0.9456	0.8074	0.5721	0.3174	0.1326	0.0403
15	1.0000	1.0000	0.9999	0.9971	0.9738	0.8849	0.6946	0.4402	0.2142	0.0769
16	1.0000	1.0000	1.0000	0.9990	0.9884	0.9367	0.7978	0.5681	0.3185	0.1341
17	1.0000	1.0000	1.0000	0.9997	0.9953	0.9680	0.8761	0.6885	0.4391	0.2148
18	1.0000	1.0000	1.0000	0.9999	0.9983	0.9852	0.9301	0.7911	0.5651	0.3179
19	1.0000	1.0000	1.0000	1.0000	0.9994	0.9937	0.9637	0.8702	0.6844	0.4373
20	1.0000	1.0000	1.0000	1.0000	0.9998	0.9976	0.9827	0.9256	0.7870	0.5627
21	1.0000	1.0000	1.0000	1.0000	1.0000	0.9991	0.9925	0.9608	0.8669	0.6821
22	1.0000	1.0000	1.0000	1.0000	1.0000	0.9997	0.9970	0.9811	0.9233	0.7852
23	1.0000	1.0000	1.0000	1.0000	1.0000	0.9999	0.9989	0.9917	0.9595	0.8659
24	1.0000	1.0000	1.0000	1.0000	1.0000	1.0000	0.9996	0.9966	0.9804	0.9231
25	1.0000	1.0000	1.0000	1.0000	1.0000	1.0000	0.9999	0.9988	0.9914	0.9597
26	1.0000	1.0000	1.0000	1.0000	1.0000	1.0000	1.0000	0.9996	0.9966	0.9808
27	1.0000	1.0000	1.0000	1.0000	1.0000	1.0000	1.0000	0.9999	0.9988	0.9917
28	1.0000	1.0000	1.0000	1.0000	1.0000	1.0000	1.0000	1.0000	0.9996	0.9968
29	1.0000	1.0000	1.0000	1.0000	1.0000	1.0000	1.0000	1.0000	0.9999	0.9989
30	1.0000	1.0000	1.0000	1.0000	1.0000	1.0000	1.0000	1.0000	1.0000	0.9997
31	1.0000	1.0000	1.0000	1.0000	1.0000	1.0000	1.0000	1.0000	1.0000	0.9999
32	1.0000	1.0000	1.0000	1.0000	1.0000	1.0000	1.0000	1.0000	1.0000	1.0000

$p =$	0.05	0.10	0.15	0.20	0.25	0.30	0.35	0.40	0.45	0.50
$n = 50, x = 0$	0.0769	0.0052	0.0003	0.0000	0.0000	0.0000	0.0000	0.0000	0.0000	0.0000
1	0.2794	0.0338	0.0029	0.0002	0.0000	0.0000	0.0000	0.0000	0.0000	0.0000
2	0.5405	0.1117	0.0142	0.0013	0.0001	0.0000	0.0000	0.0000	0.0000	0.0000
3	0.7604	0.2503	0.0460	0.0057	0.0005	0.0000	0.0000	0.0000	0.0000	0.0000
4	0.8964	0.4312	0.1121	0.0185	0.0021	0.0002	0.0000	0.0000	0.0000	0.0000
5	0.9622	0.6161	0.2194	0.0480	0.0070	0.0007	0.0001	0.0000	0.0000	0.0000
6	0.9882	0.7702	0.3613	0.1034	0.0194	0.0025	0.0002	0.0000	0.0000	0.0000
7	0.9968	0.8779	0.5188	0.1904	0.0453	0.0073	0.0008	0.0001	0.0000	0.0000
8	0.9992	0.9421	0.6681	0.3073	0.0916	0.0183	0.0025	0.0002	0.0000	0.0000
9	0.9998	0.9755	0.7911	0.4437	0.1637	0.0402	0.0067	0.0008	0.0001	0.0000
10	1.0000	0.9906	0.8801	0.5836	0.2622	0.0789	0.0160	0.0022	0.0002	0.0000
11	1.0000	0.9968	0.9372	0.7107	0.3816	0.1390	0.0342	0.0057	0.0006	0.0000
12	1.0000	0.9990	0.9699	0.8139	0.5110	0.2229	0.0661	0.0133	0.0018	0.0002
13	1.0000	0.9997	0.9868	0.8894	0.6370	0.3279	0.1163	0.0280	0.0045	0.0005
14	1.0000	0.9999	0.9947	0.9393	0.7481	0.4468	0.1878	0.0540	0.0104	0.0013
15	1.0000	1.0000	0.9981	0.9692	0.8369	0.5692	0.2801	0.0955	0.0220	0.0033
16	1.0000	1.0000	0.9993	0.9856	0.9017	0.6839	0.3889	0.1561	0.0427	0.0077
17	1.0000	1.0000	0.9998	0.9937	0.9449	0.7822	0.5060	0.2369	0.0765	0.0164
18	1.0000	1.0000	0.9999	0.9975	0.9713	0.8594	0.6216	0.3356	0.1273	0.0325
19	1.0000	1.0000	1.0000	0.9991	0.9861	0.9152	0.7264	0.4465	0.1974	0.0595
20	1.0000	1.0000	1.0000	0.9997	0.9937	0.9522	0.8139	0.5610	0.2862	0.1013
21	1.0000	1.0000	1.0000	0.9999	0.9974	0.9749	0.8813	0.6701	0.3900	0.1611
22	1.0000	1.0000	1.0000	1.0000	0.9990	0.9877	0.9290	0.7660	0.5019	0.2399
23	1.0000	1.0000	1.0000	1.0000	0.9996	0.9944	0.9604	0.8438	0.6134	0.3359
24	1.0000	1.0000	1.0000	1.0000	0.9999	0.9976	0.9793	0.9022	0.7160	0.4439
25	1.0000	1.0000	1.0000	1.0000	1.0000	0.9991	0.9900	0.9427	0.8034	0.5561
26	1.0000	1.0000	1.0000	1.0000	1.0000	0.9997	0.9955	0.9686	0.8721	0.6641
27	1.0000	1.0000	1.0000	1.0000	1.0000	0.9999	0.9981	0.9840	0.9220	0.7601
28	1.0000	1.0000	1.0000	1.0000	1.0000	1.0000	0.9993	0.9924	0.9556	0.8389
29	1.0000	1.0000	1.0000	1.0000	1.0000	1.0000	0.9997	0.9966	0.9765	0.8987
30	1.0000	1.0000	1.0000	1.0000	1.0000	1.0000	0.9999	0.9986	0.9884	0.9405
31	1.0000	1.0000	1.0000	1.0000	1.0000	1.0000	1.0000	0.9995	0.9947	0.9675
32	1.0000	1.0000	1.0000	1.0000	1.0000	1.0000	1.0000	0.9998	0.9978	0.9836
33	1.0000	1.0000	1.0000	1.0000	1.0000	1.0000	1.0000	0.9999	0.9991	0.9923
34	1.0000	1.0000	1.0000	1.0000	1.0000	1.0000	1.0000	1.0000	0.9997	0.9967
35	1.0000	1.0000	1.0000	1.0000	1.0000	1.0000	1.0000	1.0000	0.9999	0.9987
36	1.0000	1.0000	1.0000	1.0000	1.0000	1.0000	1.0000	1.0000	1.0000	0.9995
37	1.0000	1.0000	1.0000	1.0000	1.0000	1.0000	1.0000	1.0000	1.0000	0.9998
38	1.0000	1.0000	1.0000	1.0000	1.0000	1.0000	1.0000	1.0000	1.0000	1.0000

Answers

Prior knowledge 1

1 a Mean 5.89 (2 d.p.); Median 6; Mode 4; Range 10
b Mean 18.38 (2 d.p.); Median 18.5; Mode 20; Range 9
2 ANY TWO FROM: Overlapping categories; No option for > 4 hours; Question doesn't specify a period of time.
How much TV do you watch each day?

| 0–1 hours | 2–3 hours | 4 hours or more |

3 a 29   **b** 35   **c** 38   **d** 95

Exercise 1A

1 a A census observes or measures every member of a population.
b Advantage: will give a completely accurate result. Disadvantage: ANY ONE FROM: time consuming, expensive.
2 a The testing process will destroy the harness, so a census would destroy *all* the harnesses.
b 250 kg is the median load at which the harnesses in the sample break. This means that half of the harnesses will break at a load less than 250 kg.
c Test a larger number of harnesses.
3 a ANY ONE FROM:
It would be expensive.
It would be time consuming.
It would be difficult.
b A list of residents.   **c** A resident.
4 a The testing process will destroy the microswitches, so a census would destroy *all* the switches.
b The mean is less than the stated average but one of the switches lasted a significantly lower number of operations which suggests the median might be a better average to take – not affected by outliers. The data supports the company claim.
c Test a larger number of microswitches.
5 a All the mechanics in the garage.
b Everyone's views will be known.

Exercise 1B

1 a Year 1: 8, Year 2: 12, Year 3: 16
b ANY ONE FROM: sample accurately reflects the population structure of the school; guarantees proportional representation of different year groups in the sample.
2 a Patterns in the sample data might occur when taking every 20th person.
b A simple random sample using the alphabetical list as the sampling frame.
3 a No: A systematic sample requires the first selected person to be chosen at random.
b Take a simple random sample using the list of members as the sampling frame.
4 a Stratified sampling.
b Male Y12: 10, Male Y13: 7, Female Y12: 12, Female Y13: 11
5 $k = \dfrac{480}{30} = 16$
Randomly select a number between 1 and 16. Starting with the worker with this clocking-in number, select the workers that have every 16th clocking-in number after this.
6 a Any method in which every member of the population has an equal chance of being selected, e.g. lottery. Disadvantage: the sample may not accurately reflect the proportions of members at the club who play each sport.

b The sample will have proportional representation of the members who play the different sports.
c Cricket: 10, Hockey: 12, Squash: 8

Exercise 1C

1 a i Divide the population into groups according to given characteristics. The size of each group determines the proportion of the sample that should have that characteristic. The interviewer assesses which group people fall into as part of the interview. Once a quota has been filled, no more people in that group are interviewed.
  ii Opportunity sampling consists of taking the sample from the people who are available at the time the study is carried out, e.g. the first 40 shoppers who are available to be interviewed.
b Quota sampling.
2 Similarities: The population is divided according to the characteristics of the whole population (into strata for stratified sampling, and groups for quota sampling)
Differences: Stratified sampling uses random sampling whereas quota sampling does not.
3 a Opportunity sampling
b Sample is likely to be biased towards people who eat fish and chips on a Friday.
c Survey people at different times of day. Survey people in other parts of the town, not outside the fish and chip shop.
4 a 5.4 hours
b Opportunity sampling; unlikely to provide a representative sample of the town as a whole
c Increase the number of people asked. Ask people at different times/in different locations.
5 a Quota sampling.
b ANY ONE FROM: no sampling frame required, quick, easy, inexpensive, allows for comparison between male and female deer.
c Males are on average heavier and have a greater spread.
d Increase the sample size. Catch deer at random times during the day.
6 a Student's opportunity sample: For example, first five values
b 1.9, 2.0, 2.6, 2.3, 2.0
c 1.96 m, 2.16 m
d Systematic sample – is random and likely to be more representative. Opportunity sample might get all the small values, for example.

Exercise 1D

1 a Quantitative   **b** Qualitative
  c Quantitative   **d** Quantitative
  e Qualitative
2 a Discrete   **b** Continuous
  c Discrete   **d** Continuous
  e Continuous   **f** Continuous
3 a It is descriptive rather than numerical.
b It is quantitative because it is numerical. It is discrete because its value must be an integer; you cannot have fractions of a pupil.
c It is quantitative because it is numerical. It is continuous because weight can take any value in a given range.

4 a 1.4 kg and 1.5 kg   **b** 1.35 kg
   c 0.1 kg

Exercise 1E

1 a Leuchars
   b Perth
   c ANY ONE FROM: Leeming, Heathrow, Beijing
   d ANY ONE FROM: Leuchars, Hurn, Camborne, Jacksonville, Perth
   e ANY ONE FROM: Beijing, Jacksonville, Perth
2 Continuous – it can take any value in the range 0 to 100
3 a i 10.14 hours   **ii** 7.6 hours
   b i 9.5 hours   **ii** 12.8 hours
   c The mean of the daily total sunshine in Leeming is higher than that in Heathrow. Leeming is north of Heathrow, so these data do not support Supraj's conclusion.
4 0.14 mm, treat tr. as 0 in numerical calculations.
5 a i Covers several months   **ii** Small sample size
   b Two consecutive days chosen all the time – not random, possibly have similar weather.
   c Number the days and choose a simple random sample.
6 a Perth is in the southern hemisphere so August is a winter month
   b The lowest temperatures in the UK are at coastal locations (Camborne and Leuchars). The highest temperature is at an inland location (Beijing). There is some evidence to support this conclusion.
7 Oktas measure the cloud coverage in eighths. The highest value is 8 which represents full cloud coverage.
8 a She needs to select days at regular intervals in an ordered list. Put the days into date order. Select every sixth day (184 ÷ 30 = 6.13).
   b Some of the data values might not be available (n/a).

Large data set

1 a 1020 hPa
   b 0.0 mm
   c i

Temperature, t (°C)	Frequency
$10 \leqslant t < 15$	1
$15 \leqslant t < 20$	50
$20 \leqslant t < 25$	10
$25 \leqslant t < 30$	1

ii

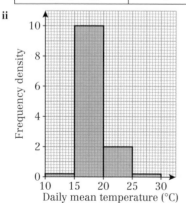

iii

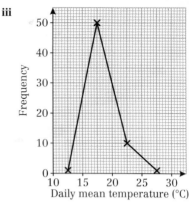

2 Students' own answer.

Mixed exercise 1

1 a 9.6°C
   b Sampling frame: first 15 days in May 1987
     Allocate each date a number from 1 to 15
     Use the random number function on calculator to generate 5 numbers between 1 and 15
   c Students' own answers.
   d 10.8°C
2 a i Advantage: very accurate; disadvantage: expensive (time consuming).
     ii Advantage: easier data collection (quick, cheap); disadvantage: possible bias.
   b Assign unique 3-digit identifiers 000, 001, ..., 499 to each member of the population. Work along rows of random number tables generating 3-digit numbers. If these correspond to an identifier then include the corresponding member in the sample; ignore repeats and numbers greater than 499. Repeat this process until the sample contains 100 members.
3 a i Collection of individual items.
     ii List of sampling units.
   b i List of registered owners from DVLC.
     ii List of people visiting a doctor's clinic in Oxford in July 1996.
4 a Advantage – the results are the most representative of the population since the structure of the sample reflects the structure of the population.
     Disadvantage – you need to know the structure of the population before you can take a stratified sample.
   b Advantage – quick and cheap.
     Disadvantage – can introduce bias (e.g. if the sample, by chance, only includes very tall people in an investigation into heights of students).
5 a People not in office not represented.
   b i Get a list of the 300 workers at the factory.
     $\frac{300}{30} = 10$ so choose one of the first ten workers on the list at random and every subsequent 10th worker on the list, e.g. if person 7 is chosen, then the sample includes workers 7, 17, 27, ..., 297.
     ii The population contains 100 office workers ($\frac{1}{3}$ of population) and 200 shop floor workers ($\frac{2}{3}$ of population).
     The sample should contain $\frac{1}{3} \times 30 = 10$ office workers and $\frac{2}{3} \times 30 = 20$ shop floor workers. The 10 office workers in the sample should be a simple random sample of the 100 office workers. The 20 shop floor workers should be a simple random sample of the 200 shop floor workers.

Online Full worked solutions are available in SolutionBank.

iii Decide the categories e.g. age, gender, office/non office and set a quota for each in proportion to their numbers in the population. Interview workers until quotas are full.

6 a Allocate a number between 1 and 120 to each pupil. Use random number tables, computer or calculator to select 15 different numbers between 1 and 120 (or equivalent).
Pupils corresponding to these numbers become the sample.

b Allocate numbers 1–64 to girls and 65–120 to boys.
Select $\frac{64}{120} \times 15 = 8$ different random numbers between 1 and 64 for girls.
Select 7 different random numbers between 65 and 120 for boys. Include the corresponding boys and girls in the sample.

7 a Stratified sampling.

b Uses naturally occurring (strata) groupings. The results are more likely to represent the views of the population since the sample reflects its structure.

8 a Opportunity sampling.

b ANY ONE FROM: Easy to carry out, Inexpensive.

c Continuous – weight can take any value.

d 76 kg

e 79.6 kg

f The second conservationist is likely to have a more reliable estimate as opportunity sampling is unlikely to provide a representative sample.

g Select more springboks at each location.

9 a Not random – the dates are selected at regular intervals so it is a systematic sample.

b Select the first date at random and then the same date each month – systematic sample. Advantage: each month covered; Disadvantage: may be patterns in the sample data. Select the six days at random – simple random sample. Advantage: avoids likelihood of patterns; Disadvantage: May not cover the full range of months.

c Continuous – rainfall can take any value.

d 8.2 mm

e This estimate is unlikely to be reliable as it does not include the winter months.

Large data set

a Student's own answer.

b Simple and quick to use.

c Student's own answer.

d The sampling frame is not random (it is in date order) so systematic sampling could introduce bias. Could improve the estimate by using a random sample.

CHAPTER 2

Prior knowledge 2

1 a Qualitative   **b** Quantitative
  c Qualitative   **d** Quantitative

2 a Discrete   **b** Continuous
  c Continuous   **d** Discrete
  e Continuous

3 Mean: 5.33, Median: 6, Mode: 6, Range: 4

Exercise 2A

1 a 700 g   **b** 600 g   **c** 700 g
  d The mean will increase; the mode will remain unchanged; the median will decrease.

2 a 42.7

b The mean will increase.

3 a May: 23 355 m, June: 21 067 m

b 22 230 m

4 a 8 minutes   **b** 10.2 minutes   **c** 8.5 minutes

d The median would be best. The mean is affected by the extreme value 26.

5 a 2   **b** 1   **c** 1.47   **d** the median

6 6.31 petals

7 1

Exercise 2B

1 a £351 to £400   **b** £345   **c** £351 to £400

2 a 82.3 decibels

b The mean is an estimate as we don't know the exact noise levels recorded.

3 a $10 < t \leq 12$

b 11.4 °C

4 Store B (mean 51 years) employs older workers than store A (mean 50 years).

Exercise 2C

1 a 1020 hPa   **b** $Q_1 = 1017$ hPa, $Q_3 = 1024.5$ hPa

2 Median 37, $Q_1 = 37$, $Q_3 = 38$

3 1.08

4 a 432 kg   **b** 389 kg   **c** 480 kg

d Three-quarters of the cows weigh 480 kg or less.

5 a 44.0 minutes   **b** 48.8 minutes

c 90th percentile = 57.8 minutes so 10% of customers have to wait longer 57.8 minutes, not 56 minutes as stated by the firm.

6 a 2.84 m. 80% of condors have a wingspan of less than 2.84 m.

b The 90th percentile is in the $3.0 \leq w$ class. There is no upper boundary for this class, so it is not possible to estimate the 90th percentile.

Exercise 2D

1 a 71   **b** 24.6   **c** 193.1 mm   **d** 7

2 a £81.87   **b** 22

3 a 6.2 minutes   **b** 54

4 a Median 11.5 °C, $Q_1 = 10.3$ °C, $Q_3 = 12.7$ °C, IQR = 2.4 °C

b On average, the temperature was higher in June than in May (higher median). The temperature was more variable in May than June (higher IQR).

c 24 days

Exercise 2E

1 a 3   **b** 0.75   **c** 0.866

2 3.11 kg

3 a 178 cm   **b** 59.9 cm²   **c** 7.74 cm

4 Mean 5.44, standard deviation 2.35

5 a Mean £10.22, standard deviation £1.35

b 19

6 1.23 days

7 Mean 16.1 hours, standard deviation 4.69 hours
One standard deviation below mean 11.41 hours.
41 parts tested (82%) lasted longer than one standard deviation below the mean. According to the manufacturers, this should be 45 parts (90%), so the claim is false.

8 a Mean 8.1 kn, standard deviation 3.41 kn

b 12 days

c The windspeeds are equally distributed throughout the range.

Exercise 2F

1 **a** 11, 9, 5, 8, 3, 7, 6   **b** 7   **c** 70
2 **a** 7, 10, 4, 10, 5, 11, 2, 3   **b** 6.5   **c** 48.5
3 365
4 2.34
5 **a** 1.2 hours   **b** 25.1 hours   **c** 1.76 hours
6 22.9
7 416 mm
8 **a** $t = 0.8(m + 12)$ or $t = \dfrac{m + 12}{1.25}$
   b Mean 54, standard deviation 0.64
9 Mean 1020 hPa, standard deviation 6.28 hPa

Mixed exercise 2

1 69.2
2 **a** 10, 12, 9, 2, 2.5, 9.5   **b** 7.5   **c** 607
3 £18 720
4 **a** Group A 63.4, group B 60.2
   b The method used for group A may be better.
5 **a** 21 to 25 hours   **b** 21.6 hours
   c 20.6 hours   **d** 20.8 hours
6 37.5
7 **a** 20.5   **b** 34.7   **c** 14.2
8 **a** 13.1
   b Variance 102, standard deviation 10.1 minutes
9 **a** 98.75 mm **b** 104 mm **c** 5.58 mm **d** 4.47 mm
10 **a** Mean 13.5, standard deviation 1.36
   b 4.0 °C   **c** 5 days
11 **a** Mean 3.42, standard deviation 1.61
   b Mean 9.84 knots, standard deviation 3.22 knots
12 **a** Mean 15.8 cm, standard deviation 2.06 cm
   b The mean wingspan will decrease.
   c Mean 57 cm, standard deviation 3 cm

Challenge

Mean 3.145 cm, standard deviation 1.39 cm

CHAPTER 3

Prior knowledge 3

1 **a**
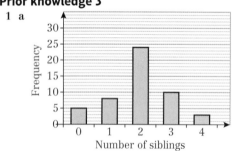

   b
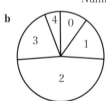

2 11
3 Mean 28.5, standard deviation 7.02

Exercise 3A

1 **a** 7 is an outlier   **b** 88 is not an outlier
   c 105 is an outlier
2 **a** No outliers   **b** 170 g and 440 g
   c 760 g

3 **a** 11.5 kg
   b Smallest 2.0 kg, largest 10.2 kg
4 **a** Mean 10.2, standard deviation 7.36
   b It is an outlier as it is more than 2 standard
     deviations above the mean.
   c e.g. It could be the age of a parent at the party.
   d Mean 7.75, standard deviation 2.44

Exercise 3B

1
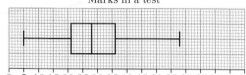

2 **a** 47, 32   **b** 38   **c** 15   **d** 64
3 **a** The male turtles have a higher median mass,
     a greater interquartile range and a greater total
     range.
   b It is more likely to have been female. Very few of
     the male turtles had a mass this low, but more than
     a quarter of the female turtles had a mass of more
     than this.
   c 500 g
4 **a** $Q_1 = 22$ knots, Q_2 26 knots, $Q_3 = 30$ knots
   b IQR = 8
     1.5 × IQR above Q_3 = 42
     46 > 42 and 78 > 42, so 46 and 78 are outliers.
   c
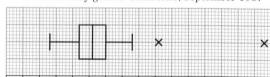

Exercise 3C

1 **a**

Cumulative frequency graph with axes Cumulative frequency (0 to 140) against Mass, m (kg) (0.5 to 2.5)

   b ≈ 1.6 kg
   c IQR ≈ 0.3, 10th to 90th interpercentile
     range ≈ 0.65

d

Masses of Coulter pine cones

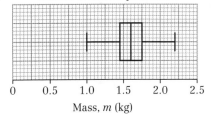

Mass, m (kg)

2 a

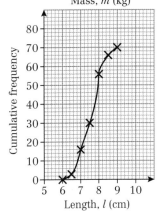

Length, l (cm)

b Median ≈ 7.6, Q_1 ≈ 7.1, Q_3 ≈ 8
c i ≈ 8   **ii** ≈ 24
d

Length of earthworms

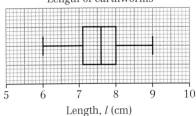

Length, l (cm)

3 a

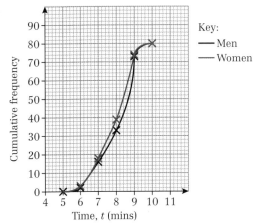

Time, t (mins)

b Women
c Men
d Men ≈ 24, women ≈ 28

4 a

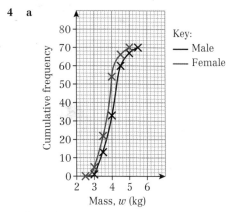

Mass, w (kg)

b Male   **c** Female

Exercise 3D

1

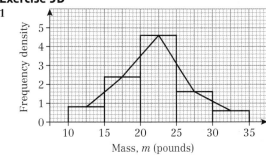

Mass, m (pounds)

2 a The quantity (time) is continuous.
  b 150   **c** 369   **d** 699
3 a The quantity (distance) is continuous.
  b 310   **c** 75   **d** 95   **e** 65
4 a 32 lambs is represented by 100 small squares,
     therefore 25 small squares represents 8 lambs.
  b 32   **c** 168   **d** 88
5 a i

Time, t (min)	Frequency
$0 \leqslant t < 20$	4
$20 \leqslant t < 30$	10
$30 \leqslant t < 35$	15
$35 \leqslant t < 40$	25
$40 \leqslant t < 50$	7
$50 \leqslant t < 70$	6

  ii

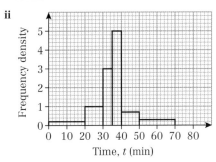

Time, t (min)

  b 35
6 a 12.5 and 14.5
  b i 6 cm   **ii** 3 cm
7 a Width 0.5 cm, height 14 cm
  b Mean 10.4, standard deviation 2.4
  c 9 °C   **d** 4.7 days

Exercise 3E

1 The median speed is higher on motorway A than on motorway B. The spread of speeds for motorway B is greater than the spread of speeds for motorway A (comparing IQRs).

2 Class 2B: mean 32.5, standard deviation 6.6
Class 2F: mean 27.2, standard deviation 11.4
The mean time for Class 2B is higher than the mean time for Class 2F. The standard deviation for Class 2F is bigger than for Class 2B, showing that the times were more spread out.

3 The median height for boys (163 cm) is higher than the median height for girls (158 cm). The spread of heights for boys is greater than the spread of heights for girls (comparing IQRs).

4 **a** Leuchars: median 100, $Q_1 = 98$, $Q_3 = 100$
   Camborne: median 98, $Q_1 = 92$, $Q_3 = 100$

   b The median humidity in Leuchars is higher than the median humidity in Camborne. The spread of humidities for Camborne is greater than the spread of humidities for Leuchars.

Large data set

1 **a** 1987: 6.6 kn, 2015: 7.7 kn
   b 1987: 4 kn, 2015: 7 kn
   c 1987: 3.0 kn, 2015: 2.8 kn

2 The mean windspeeds were higher in 2015 than in 1987. The spread of the speeds was greater in 1987 than in 2015 (higher standard deviation).

Mixed exercise 3

1 **a** $Q_1 = 178$, $Q_2 = 185$, $Q_3 = 196$
   b 226
   c

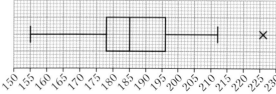

Distance travelled each day

Distance (km)

2 **a** 45 minutes   **b** 60 minutes
   c This represents an outlier.
   d Irt has a higher median than Esk.
   The interquartile ranges were about the same.
   e Esk had the fastest runners.
   f Advantages: easy to compare quartiles, median and spread. Disadvantages: cannot compare mean or mode.

3 **a**

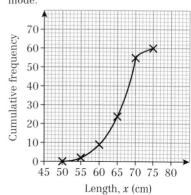

Length, x (cm)

b ≈ 66 cm
c ≈ 6.5 cm
d The distributions have very similar medians and quartiles. Maximum length of the European badgers is greater than the maximum length of the honey badgers.
e Do not have exact data values so cannot compare the median, quartiles or range accurately.

4 **a** 26
   b 17

5 **a** width = 1.5 cm, height = 2.6 cm
   b width = 7.5 cm, height = 0.28 cm

6 **a**

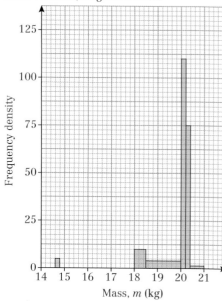

Mass, m (kg)

b Mean 19.8 kg, standard deviation 0.963 kg
c 20.1 kg

7 **a** 22.3
   b Median 20; quartiles 13, 31
   c No outliers.
   d Bags of potato crisps sold each day

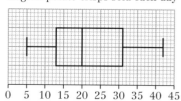

Number of bags of potato crisps sold

8 **a** The maximum gust is continuous data and the data is given in a grouped frequency table.
   b 1 cm wide and 13.5 cm tall
   c Mean 23.4, standard deviation 7.32
   d 44 days

9 **a** 1987: 11.9 °C, 2015: 12.1 °C
   b The mean temperature was slightly higher in 2015 than in 1987. The standard deviation of temperatures was higher in 1987 (2.46 °C) than in 2015 showing that the temperatures were more spread out.
   c 15 days assuming that the temperatures are equally distributed throughout the range.

Challenge:
0.6 cm

CHAPTER 4

Prior knowledge 4

1

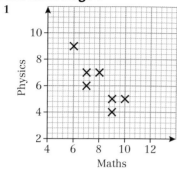

2 a −3.21
   b 0.34

Exercise 4A

1 a Positive correlation.
   b The longer the treatment, the greater the loss of weight.
2 a No correlation.
   b The scatter graph does not support the statement that hotter cities have less rainfall.
3 a

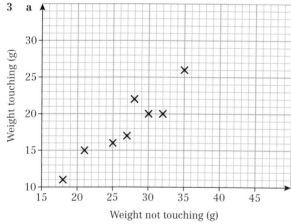

   b There is positive correlation. If a student guessed a greater weight before touching the bag, they were more likely to guess a greater weight after touching it.
4 a

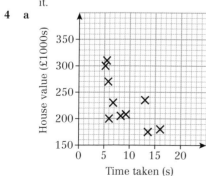

   b Weak negative correlation.
   c For example, there may be a third variable that influences both house value and internet connection, such as distance from built up areas.
5 a $Q_3 + 1.5 \times IQR = 4.85 + 7.125 = 11.975$
      $21.7 > 11.975$, therefore is an outlier.

b i There is no reason to believe that the data collected by the Met Office is incorrect.
   ii 21.7 is an outlier so may not be representative of the typical rainfall.
c

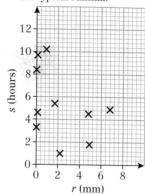

d Weak negative correlation.
e For example, there could be a causal relationship as days with more rainfall will have more clouds, and therefore less sunshine.

Exercise 4B

1 a, b

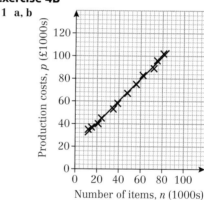

   c If the number of items produced per month is zero, the production costs will be approximately £21 000. If the number of items per month increases by 1000 items, the production costs increase by approximately £980.
   d The prediction for 74 000 is within the range of the data (interpolation) so is more likely to be accurate. The prediction for 95 000 is outside the range of the data (extrapolation) so is less likely to be accurate.

2 a

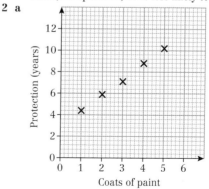

b A gradient of 1.45 means that for every extra coat of paint, the protection will increase by 1.45 years, therefore if 10 coats of paint are applied, the protection will be 14.5 years longer than if zero coats of paint were applied. After 10 coats of paint, the protection will last 2.93 + 14.5 = 17.43 years.

3 a

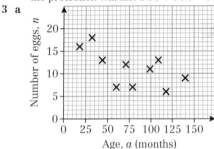

b The scatter diagram shows negative correlation, therefore the gradient in the regression equation should be negative.

4 This is not sensible as there are unlikely to be any houses with no bedrooms.

5 a For each percent increase in daily maximum relative humidity there is a decrease of 106 Dm in daily mean visibility.

b High levels of relative humidity cause mist or fog which will decrease visibility. Hence there is likely to be a causal relationship.

c i The prediction for 100% is outside the range of the data (extrapolation) so is less likely to be accurate.

ii The regression equation should only be used to predict a value for v given h.

d Data is only useful for analysing the first two weeks of September. Random values throughout September should be used and analysis made of the whole month. The sample size could also be increased across multiple months as data between May and October is available.

Mixed exercise 4

1 The data shows that the number of serious road accidents in a week strongly correlates with the number of fast food restaurants. However, it does not show whether the relationship is causal. Both variables could correlate with a third variable, e.g. the number of roads coming into a town.

2 a

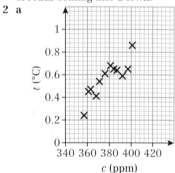

b Strong positive correlation.

c As mean CO_2 concentration in the atmosphere increases, mean global temperatures also increase.

3 a Strong positive correlation.

b If the number of items increases by 1, the time taken increases by approximately 2.64 minutes.

4 (1) 3500 is outside the range of the data (extrapolation).
(2) The regression equation should only be used to predict a value of GNP (y) given energy consumption (x).

5 a Mean + 2SD = 15.2 + 2 × 11.4 = 38; 50 > 38

b The outlier should be omitted as it is very unlikely that the average temperature was 50 °C.

c If the temperature increases by approximately 1 °C, the number of pairs of gloves sold each month decreases by 0.18.

6 a 44 is the length in centimetres of the spring with no mass attached. If a mass of 1 g is attached, the spring would increase in length by approximately 0.2 cm.

b i Outside the range of the data (extrapolation)

ii The regression equation should only be used to predict a value of s given m

7 a & b

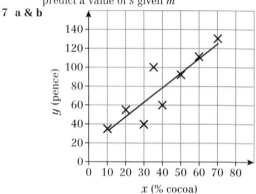

c Brand D is overpriced, since it is a long way above the line.

d The regression equation should be used to predict a value for y given x so the student's method is valid.

Large data set

1 a

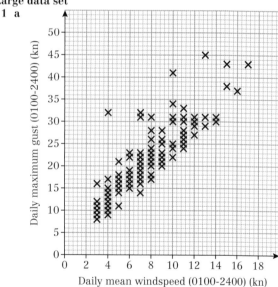

b Moderate positive correlation.

c The relationship is causal as the maximum gust is related to the mean windspeed.

d i 6.05 **ii** 15.7 **iii** 30.8 **iv** 91.0

e Parts **ii** and **iii** are within the range of the data (interpolation) so are more likely to be accurate. Parts **i** and **iv** are outside the range of the data (extrapolation) so are less likely to be accurate.

f $w = 0.053 + 0.35g$; 10.6 knots

2 a Regression equation: $s = 15.1 - 1.8c$
Estimated missing values: 1.8, 8.5, 6.3, 11.6, 3.0, 7.1, 12.2, 14.4, 9.5, 6.3, 2.0, 3.9, 7.2, 3.1, 3.7, 3.9, 0.9
b The relationship is causal because daily sunshine is related to daily mean cloud cover.

CHAPTER 5

Prior knowledge 5

1 a $\frac{2}{9}$ **b** $\frac{4}{9}$ **c** $\frac{2}{3}$ **d** 0
2 HHH, HHT, HTH, HTT, THH, THT, TTH, TTT
3 a $\frac{25}{216}$
   b $\frac{11}{36}$
   c $\frac{125}{216}$

Exercise 5A

1 $\frac{1}{2}$
2 a

		Second roll				
		1	**2**	**3**	**4**	**5**	**6**
First roll	**1**	1	2	3	4	5	6
	2	2	4	6	8	10	12
	3	3	6	9	12	15	18
	4	4	8	12	16	20	24
	5	5	10	15	20	25	30
	6	6	12	18	24	30	36

   b i $\frac{1}{18}$ **ii** $\frac{2}{9}$ **iii** $\frac{3}{4}$
3 a $\frac{2}{5}$
   b $\frac{5}{7}$
   c Less likely; frequency uniformly distributed throughout the class.
4 a $\frac{19}{40}$ **b** $\frac{109}{240}$ **c** $\frac{71}{240}$
   d $\frac{2}{15}$; distribution of lengths of koalas between 70 and 75 cm is uniform.
5 a $\frac{16}{35}$
   b $\frac{32}{35}$

Challenge:
5, 7 or 9

Exercise 5B

1 a

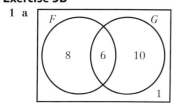

   i $\frac{14}{25}$ **ii** $\frac{6}{25}$ **iii** $\frac{8}{25}$ **iv** $\frac{1}{25}$

2 a

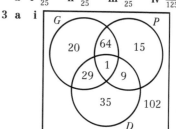

   b i $\frac{3}{25}$ **ii** $\frac{2}{25}$ **iii** $\frac{2}{25}$ **iv** $\frac{54}{125}$
3 a i

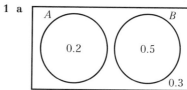

   b i $\frac{89}{275}$ **ii** $\frac{103}{275}$ **iii** $\frac{14}{55}$ **iv** $\frac{102}{275}$
4 a 0.17
   b 0.18
   c 0.55
5 a 0.3
   b 0.3
6 a 0.15
   b 0.15
7 $p = 0.13, q = 0.25$

Challenge
$p = 0.115, q = 0.365, r = 0.12$

Exercise 5C

1 a

   b 0.7   **c** 0.3
2 P(sum of 4) + P(same number) ≠ P(sum of 4 or same number), so the events are not mutually exclusive.
3 0.15
4 0.3
5 a Bricks and trains; their curves do not overlap.
   b Not independent.
6 a 0.25   **b** Not independent
7 a P(S and T) = 0.3 – 0.18 = 0.12
     P(S) × P(T) = 0.3 × 0.4 = 0.12 = P(S and T)
     So S and T are independent.
   b i 0.12   **ii** 0.42
8 P(W) × P(X) = 0.5 × 0.45 = 0.225
   P(W and X) = 0.25, so W and X are **not** independent.
9 a $x = 0.15, y = 0.3$
   b P(F and R) = 0.15 ≠ P(F) × P(R) = 0.45 × 0.4 = 0.18
10 $p = 0.14$ and $q = 0.33$ or $p = 0.33$ and $q = 0.14$

Challenge
a Set P(A) = p and P(B) = q, then P(A and B) = pq
   P(A and not B) = P(A) – P(A and B) = $p - pq$
   P(not B) = $1 - q$
   $\Rightarrow$ P(A) × P(not B) = $p(1 - q) = p - pq$ = P(A and not B)

b P(not A and not B) = 1 − P(A or B)
    = 1 − P(A) − P(B) + P(A and B)
    = 1 − p − q + pq = (1 − p)(1 − q)
  But P(not A) = 1 − p and P(not B) = 1 − q, so
  P(not A and not B) = P(not A) × P(not B)

Exercise 5D

1 a

Bead 1   Bead 2

$\frac{3}{8}$ Red
$\frac{3}{8}$ Red
    $\frac{5}{8}$ Blue
$\frac{5}{8}$ Blue
    $\frac{3}{8}$ Red
    $\frac{5}{8}$ Blue

b $\frac{25}{64}$ **c** $\frac{5}{8}$

2 a

$\frac{5}{9}$ Odd
    $\frac{1}{2}$ Odd
    $\frac{1}{2}$ Even
$\frac{4}{9}$ Even
    $\frac{5}{8}$ Odd
    $\frac{3}{8}$ Even

b $\frac{1}{6}$ **c** $\frac{5}{9}$

3 a

0.4 Bus
    0.2 Late
    0.8 On time
0.6 Walk
    0.3 Late
    0.7 On time

b 0.26

4 a

0.7 Par or under
    0.8 Par or under
    0.2 Over par
0.3 Over par
    0.4 Par or under
    0.6 Over par

b Not independent   **c** 0.26

5 a

$\frac{1}{3}$ Heads
    $\frac{1}{3}$ Heads
        $\frac{1}{3}$ Heads
        $\frac{2}{3}$ Tails
    $\frac{2}{3}$ Tails
        $\frac{1}{3}$ Heads
        $\frac{2}{3}$ Tails
$\frac{2}{3}$ Tails
    $\frac{1}{3}$ Heads
        $\frac{1}{3}$ Heads
        $\frac{2}{3}$ Tails
    $\frac{2}{3}$ Tails
        $\frac{1}{3}$ Heads
        $\frac{2}{3}$ Tails

b $\frac{1}{27}$ **c** $\frac{4}{9}$ **d** $\frac{1}{9}$

6 a $\frac{5}{26}$ **b** $\frac{4}{11}$ **c** $\frac{36}{143}$

Mixed exercise 5

1 a $\frac{392}{3375}$ **b** $\frac{14}{75}$

2 a 0.0397   **b** 0.286   **c** 0.714

3 a $\frac{64}{125}$

b $\frac{8}{25}$

c $\frac{33}{250}$

d $\frac{74}{125}$, using interpolation and assuming uniform distribution of scores

4 a $\frac{44}{50}$ **b** $\frac{77}{100}$

5 a

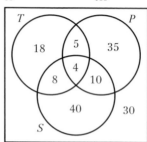

b i 0.2   **ii** 0.82

6 a

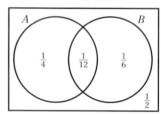

b P(A) = $\frac{1}{3}$, P(B) = $\frac{1}{4}$, P(A and B) = $\frac{1}{12}$
  P(A) × P(B) = P(A and B), so A and B are independent.

7 a Cricket and swimming
b Not independent

8 a

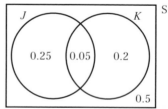

b P(J) = 0.3, P(K) = 0.25, P(J and K) = 0.05
  P(J) × P(K) = 0.075 ≠ P(J and K), so J and K are not independent.

9 a 0.5
b

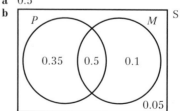

c 0.35
d No. P(P) = 0.85 and P(M) = 0.6, so
  P(P) × P(M) = 0.51 ≠ P(P and M)

10 Not independent

Online Full worked solutions are available in SolutionBank.

11 a

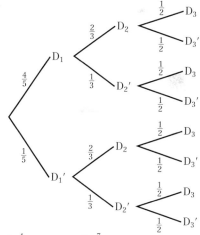

b **i** $\frac{4}{15}$ **ii** $\frac{7}{30}$

c $\frac{11}{15}$

12 a

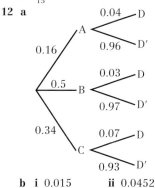

b **i** 0.015   **ii** 0.0452

Challenge
0.2016

CHAPTER 6

Prior knowledge 6

1 a $\frac{1}{8}$ **b** $\frac{1}{8}$ **c** $\frac{3}{8}$ **d** $\frac{1}{2}$

2 a $\frac{4}{36}$ **b** $\frac{18}{36}$ **c** $\frac{18}{36}$ **d** $\frac{12}{36}$ **e** $\frac{15}{36}$

Exercise 6A

1 a This is not a discrete random variable, since height is continuous quantity.

b This is a discrete random variable, since it is always a whole number and it can vary.

c This is not a discrete random variable, since the number of days in a given week is always 7.

2 0, 1, 2, 3, 4

3 a (2, 2) (2, 3) (3, 2) (3, 3)

b **i**

x	4	5	6
P($X = x$)	0.25	0.5	0.25

ii
$$P(X = x) = \begin{cases} 0.25, x = 4, 6 \\ 0.5, x = 5 \end{cases}$$

4 $\frac{1}{12}$

5 $k + 2k + 3k + 4k = 1$,
so $10k = 1$, so $k = \frac{1}{10}$.

6 a 0.125   **b** 0.875

7 a 0.3

b

x	−2	−1	0	1	2
P($X = x$)	0.1	0.1	0.3	0.3	0.2

c 0.7

8 0.25

9 a 0.02   **b** 0.46   **c** 0.56

10 a 0.625   **b** 0.375   **c** 0

11 a

s	1	2	3	4
P($S = s$)	$\frac{2}{3}$	$\frac{2}{9}$	$\frac{2}{27}$	$\frac{1}{27}$

b $\frac{1}{9}$

12 a

x	P($X = x$)
0	0.07776
1	0.2592
2	0.3456
3	0.2304
4	0.0768
5	0.01024

b

y	P($X = y$)
0	0.32768
1	0.4096
2	0.2048
3	0.0512
4	0.0064
5	0.00032

c

z	P($X = z$)
1	0.4
2	0.24
3	0.144
4	0.0864
5	0.1296

13 a The sum of the probabilities is not 1.

b $2\frac{22}{61}$

Challenge
0.625

Exercise 6B

1 a 0.273   **b** 0.0683   **c** 0.195

2 a 0.00670   **b** 0.214   **c** 0.00178

3 a $X \sim B(20, 0.01)$, $n = 20$, $p = 0.01$
Assume bolts being defective are independent of each other.

b $X \sim B(6, 0.52)$, $n = 6$, $p = 0.52$
Assume the lights operate independently and the time lights are on/off is constant.

c $X \sim B(30, \frac{1}{8})$, $n = 30$, $p = \frac{1}{8}$
Assume serves are independent and probability of an ace is constant.

4 a $X \sim B(14, 0.15)$ is OK if we assume the children in the class being Rh⁻ is independent from child to child (so no siblings/twins).

b This is not binomial since the number of tosses is not fixed. The probability of a head at each toss is constant ($p = 0.5$) but there is no value of n.

c Assuming the colours of the cars are independent (which should be reasonable).
X = number of red cars out of 15
$X \sim B(15, 0.12)$

5 a 0.358   **b** 0.189

6 a The random variable can take two values, faulty or not faulty.
There are a fixed number of trials, 10, and fixed probability of success: 0.08.
Assuming each member in the sample is independent, a suitable model is $X \sim B(10, 0.08)$

b 0.00522

7 a Assumptions: There is a fixed sample size, there are only two outcomes for the genetic marker (i.e. fully present or not present), there is a fixed probability of people having the marker.

b 0.0108

8 a The random variable can take two values, 6 or not 6. There are a fixed number of trials (15) and a fixed probability of success (0.3), Each roll of the dice is independent. A suitable distribution is $X \sim B(15, 0.3)$

b 0.219   **c** 0.127

Exercise 6C

1 a 0.9804   **b** 0.7382   **c** 0.5638   **d** 0.3020
2 a 0.9468   **b** 0.5834   **c** 0.1272   **d** 0.5989
3 a 0.5888   **b** 0.7662   **c** 0.1442   **d** 0.2302
4 a 0.8882   **b** 0.7992   **c** 0.0599   **d** 0.1258
5 a 0.0039   **b** 0.9648   **c** 0.3633
6 a 0.2252   **b** 0.4613   **c** 0.7073
7 a $k = 13$ **b** $r = 28$
8 a $k = 1$ **b** $r = 9$ **c** 0.9801
9 a $X \sim B(10, 0.30)$ Assumptions: The random variable can take two values (listen or don't listen), there are a fixed number of trials (10) and a fixed probability of success (0.3), each member in the sample is independent.

b 0.1503   **c** $s = 8$

10 a 0.2794   **b** 0.0378   **c** $d = 5$

Mixed exercise 6

1 a

x	$P(X = x)$
1	$\frac{1}{21}$
2	$\frac{2}{21}$
3	$\frac{3}{21}$
4	$\frac{4}{21}$
5	$\frac{5}{21}$
6	$\frac{6}{21}$

b $\frac{12}{21}$

2 a 0.2   **b** 0.7

3 a

x	1	2	3	4
$P(X = x)$	0.0769	0.1923	0.3077	0.4231

b $\frac{19}{26}$

4 a The probabilities must be the same.

b i 0.0625   **ii** 0.375   **iii** 0.5

5 a 15

b

y	1	2	3	4	5
$P(X = y)$	$\frac{1}{15}$	$\frac{2}{15}$	$\frac{3}{15}$	$\frac{4}{15}$	$\frac{5}{15}$

c $\frac{9}{15}$ or $\frac{3}{5}$

6 a

t	0	1	2	3	4
$P(T = t)$	0.316	0.422	0.211	0.0469	0.00391

b 0.949

c

s	1	2	3	4	5
$P(S = s)$	0.25	0.188	0.141	0.105	0.316

d 0.562

7 a 0.114   **b** 0.0005799
c 0.9373

8 a 0.0439 or $\frac{32}{729}$ **b** 0.273

9 a 0.014 (3 d.p.)   **b** 0.747 (3 d.p.)

10 a 1 There are n independent trials.
2 n is a fixed number.
3 The outcome of each trial is success or failure.
4 The probability of success at each trial is constant.
5 The outcome of any trial is independant of any other trial.

b 0.0861   **c** $n = 90$

11 a 0.000977   **b** 0.0547
12 a 0.0531   **b** 0.243
13 a $X \sim B(10, 0.15)$
b 0.0099   **c** 0.2759
14 a 0.8692   **b** 0.0727
15 a 0.8725   **b** 0.01027   **c** 0.0002407

Challenge
0.001244

CHAPTER 7

Prior knowledge 7

1 a 0.075   **b** 0.117
c 0.0036   **d** 0.00000504
2 a $X \sim B(8, \frac{1}{6})$ **b i** 0.260   **ii** 0.0307

Exercise 7A

1 a A hypothesis is a statement made about the value of a population parameter. A hypothesis test uses a sample or an experiment to determine whether or not to reject the hypothesis.

b The null hypothesis (H_0) is what we assume to be correct and the alternative hypothesis (H_1) tells us about the parameter if our assumption is shown to be wrong.

c The test statistic is used to test the hypothesis. It could be the result of the experiment or statistics calculated from a sample.

2 a One-tailed test
b Two-tailed test
c One-tailed test

3 a The test statistic is N – the number of sixes.
b $H_0: p = \frac{1}{6}$ **c** $H_1: p > \frac{1}{6}$

4 a Shell is describing what her experiment wants to test rather than the test statistic. The test statistic is the proportion of times you get a head.
b $H_0: p = \frac{1}{2}$ **c** $H_1: p \neq \frac{1}{2}$

5 a A suitable test statistic is p – the proportion of faulty articles in a batch.

b H_0: $p = 0.1$, H_1: $p < 0.1$

c If the probability of the proportion being 0.08 or less is 5% or less the null hypothesis is rejected.

6 a A suitable test statistic is p – the proportion of people that support the candidate.

b H_0: $p = 0.55$, H_1: $p < 0.55$

c If the probability of the proportion being $\frac{7}{20}$ is 2% or more, the null hypothesis is accepted

Exercise 7B

1 a The critical value is the first value to fall inside of the critical region.

b A critical region is a region of the probability distribution which, if the test statistic falls within it, would cause you to reject the null hypothesis.

c The acceptance region is the area in which we accept the null hypothesis.

2 The critical value is $x = 5$ and the critical region is $X \geqslant 5$ since $P(X \geqslant 5) = 0.0328 < 0.05$.

3 The critical value is $x = 0$ and the critical region is $X = 0$.

4 a The critical region is $X \geqslant 13$ and $X \leqslant 3$.

b $0.037 = 3.7\%$

5 The critical value is $x = 0$. The critical region is $X = 0$.

6 a The critical region is $X = 0$ and $7 \leqslant X \leqslant 10$.

b 0.085

7 a The number of times the sample fails.

b H_0: $p = 0.3$, H_1: $p < 0.3$

c The critical value is $x = 10$ and the critical region is $X \geqslant 10$

d 4.8%

8 a The number of seedlings that survive.
H_0: $p = \frac{1}{3}$, H_1: $p > \frac{1}{3}$

b The critical value is $x = 17$ and the critical region is $X \geqslant 17$

c 5.84%

9 a H_0: $p = 0.2$, H_1: $p \neq 0.2$

b The critical region is $X \leqslant 1$ and $X \geqslant 10$

c 4.47%

Challenge

a The critical region is $X \leqslant 29$ and $X \geqslant 41$

b Chance of one observation falling within critical region = 8.8%.
Chance of two observations falling within critical region = 0.77%.

Exercise 7C

1 $0.0781 > 0.05$
There is insufficient evidence to reject H_0.

2 $0.0464 < 0.05$
There is sufficient evidence to reject H_0 so $p < 0.04$.

3 $0.0480 < 0.05$
There is sufficient evidence to reject H_0 so $p > 0.30$.

4 $0.0049 < 0.01$
There is sufficient evidence to reject H_0 so $p < 0.45$.

5 $0.0526 > 0.05$
There is insufficient evidence to reject H_0 so there is no reason to doubt $p = 0.28$.

6 $0.0020 < 0.05$
There is sufficient evidence to reject H_0 so $p > 0.32$.

7 $0.3813 > 0.05$
There is insufficient evidence to reject H_0 (not significant).
There is no evidence that the probability is less than $\frac{1}{6}$.
There is no evidence that the dice is biased.

8 a Distribution B(n, 0.68).
Fixed number of trials.
Outcomes of trials are independent.
There are two outcomes success and failure.
The probability of success is constant.

b $P(X \leqslant 3) = 0.0155 < 0.05$. There is sufficient evidence to reject the null hypothesis so $p < 0.68$. The treatment is not as effective as claimed.

9 a Critical region is $X \geqslant 13$

b 14 lies in the critical region, so we can reject the null hypothesis. There is evidence that the new technique has improved the number of plants that germinate.

10 a The number of people who support the candidate.
H_0: $p = 0.35$, H_1: $p > 0.35$

b Critical region is $X \geqslant 24$

c 28 lies in the critical region, so we can reject the null hypothesis. There is evidence that the candidate's level of popularity has increased.

Exercise 7D

1 $P(X \leqslant 10) = (0.0494 > 0.025$ (two-tailed)
There is insufficient evidence to reject H_0 so there is no reason to doubt $p = 0.5$

2 $P(X \geqslant 10) = 0.189 > 0.05$ (two tailed)
There is insufficient evidence to reject H_0 so there is no reason to doubt $p = 0.3$

3 $(X \geqslant 9) = 0.244 > 0.025$ (two-tailed)
There is insufficient evidence to reject H_0 so there is no reason to doubt $p = 0.75$

4 $P(X \leqslant 1) = 0.00000034 < 0.005$ (two-tailed)
$X = 1$ lies within the critical region, so we can reject the null hypothesis.

5 $P(X \geqslant 4) = 0.0178 > 0.01$ (two-tailed)
There is insufficient evidence to reject H_0 so there is no reason to doubt $p = 0.02$

6 $P(X \leqslant 6) = 0.0577 > 0.025$ (two-tailed)
$X = 6$ does not lie in the critical region, so there is no reason to think that the coin is biased.

7 a Critical region $X = 0$ and $X \geqslant 8$

b 4.36%

c H_0: $p = 0.2$, H_1: $p \neq 0.2$
$X = 8$ is in the critical region. There is enough evidence to reject H_0. The hospital's proportion of complications differs from the national figure.

8 Test statistic: the number of cracked bowls.
H_0: $p = 0.1$, H_1: $p \neq 0.1$
$P(X \leqslant 1) = 0.3917 = 39.17\%$
$39.17\% > 5\%$ (two-tailed) so there is not enough evidence to reject H_0. The proportion of cracked bowls has not changed.

9 Test statistic: the number of carrots longer than 7 cm
H_0: $p = 0.25$, H_1: $p \neq 0.25$
$P(X \geqslant 13) = 1 - P(X \leqslant 12) = 0.0216 = 2.16\%$
$2.16\% < 2.5\%$ (two-tailed) so there is enough evidence to reject the null hypothesis. The probability of a carrot being longer than 7 cm has increased.

10 Test statistic: the number of patients correctly diagnosed.
H_0: $p = 0.96$, H_1: $p \neq 0.96$
$P(X \leqslant 63) = 0.0000417 < 0.05$ (two-tailed) so there is enough evidence to reject the null hypothesis. The new test does not have the same probability of success as the old test.

Mixed exercise 7

1 $H_0: p = 0.2$, $H_1: p > 0.2$, $P(X \geq 3) = 0.3222 > 0.05$
There is insufficient evidence to reject H_0.
There is no evidence that the trains are late more often.

2 $H_0: p = 0.5$, $H_1: p > 0.5$, $P(X \geq 4) = 0.1875 > 0.05$
There is insufficient evidence to reject H_0.
There is insufficient evidence that the manufacturer's claim is true.

3 a Fixed number; independent trials; two outcomes (pass or fail); p constant for each car.
  b 0.16807
  c $0.3828 > 0.05$
   There is insufficient evidence to reject H_0.
   There is no evidence that the garage fails fewer than the national average.

4 a Critical region $X \leq 1$ and $X \geq 10$.
  b 0.0583
  c $H_0: p = 0.1$, $H_1: p > 0.1$, $P(X \geq 4) = 0.133 > 0.1$.
   Accept H_0. There is no evidence that the proportion of defective articles has increased.

5 $H_0: p = 0.5$, $H_1: p \neq 0.5$, $P(X \leq 8) = 0.252 > 0.025$ (two-tailed)
There is insufficient evidence to reject H_0.
There is no evidence that the claim is wrong.

6 a Critical region is $X \leq 4$ and $X \geq 16$
  b 0.0493
  c There is insufficient evidence to reject H_0.
   There is no evidence to suggest that the proportion of people buying that certain make of computer differs from 0.2.

7 a i The theory, methods, and practice of testing a hypothesis by comparing it with the null hypothesis.
   ii The critical value is the first value to fall inside of the critical region.
   iii The acceptance region is the region where we accept the null hypothesis.
  b Critical region $X = 0$ and $X \geq 8$
  c 4.36%
  d As 7 does not lie in the critical region, H_0 is not rejected. Therefore, the proportion of times that Johan is late for school has not changed.

8 $P(X \geq 21) = 0.021 < 0.05$. Therefore there is sufficient evidence to support Poppy's claim that the likelihood of a rain-free day has increased.

9 a Critical region $X \leq 5$ and $X \geq 16$
  b 5.34%
  c $X = 4$ is in the critical region so there is enough evidence to reject H_0.

10 a $X \sim B(20, 0.85)$
  b 0.1821
  c Test statistic is proportion of patients who recover.
   $H_0: p = 0.85$, $H_1: p < 0.85$
   $P(X \leq 20) = 0.00966$
   $0.00966 < 0.05$ so there is enough evidence to reject H_0. The percentage of patients who recover after treatment with the new ointment is lower than 85%.

Large data set

1 a The critical region is $X \geq 5$
  b Students' answers
  c Students' answers
2 Students' answers

Review exercise 1

1 a A census observes every member of a population. Disadvantage it would be time-consuming to get opinions from all the employees.
  b Opportunity sampling
  c Only cleaners – no managers i.e. not all types. Not a random sample – 1st 50 may be in same shift/group/share same views.
  d i Label employees (1–550) or obtain an ordered list. Select first person using random numbers (from 1–11). Then select every 11th person from the list e.g. If person 8 is selected then the sample is 8, 19, 30, 41, …
   ii Label managers (1–55) and cleaners (1–495). Use random numbers to select 5 managers and 45 cleaners.

2 a Opportunity sampling is using a sample that is available at the time the study is carried out. It is unlikely to provide a representative sample of the weather in May.
  b 87.4
  c Relative humidity above 95% gives rise to misty conditions. 4 out of 5 observations are not misty days, so she may be right. However, 5 days is not a representative sample for the whole of May.

3 a Median 27.3 miles
  b Mean 30.1 miles, standard deviation 16.6 miles

4 Mean 3.06 hours, standard deviation 3.32 hours

5 a i 37 minutes
   ii upper quartile, third quartile, 75 percentile
  b Outliers – values that are much greater than or much less than the other values and need to be treated with caution.
  c

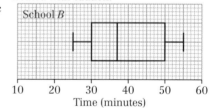

  d The children from school A generally took less time than those from school B. The median for A is less than the median for B. A has outliers, but B does not. The interquartile range for A is less than the interquartile range for B. The total range for A is greater than the total range for B.

6 a Missing frequencies: 35, 15; Missing frequency densities: 4, 6
  b 0.4   **c** 18.9 minutes
  d 7.26 minutes   **e** 18 minutes

7 0.82

8 a 4.9
  b

Online Full worked solutions are available in SolutionBank.

c Interpolation likely to be more accurate.
d 25 days

9 a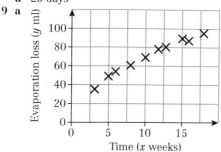

b The points lie close to a straight line.
c 3.90 ml of the chemicals evaporate each week.
d The estimate for 19 weeks is reasonably reliable, since it is just outside the range of the data. The estimate for 35 weeks is unreliable, since it is far outside the range of the data.
10 a $15.3 + 2 \times 10.2 = 35.7$ so 45 is an outlier
b A temperature of 45 °C is very high so it is likely this value was recorded incorrectly.
c When the temperature increases by 1 °C, the number of ice creams sold per month increases by 2810.
d Outside the range of the data (extrapolation)
11 a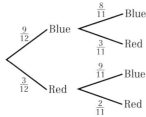

b 0.25 **c** 0.409
12 a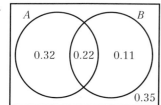

b $P(A) = 0.54$, $P(B) = 0.33$
c They are not independent.
13 a

0.85 Goodbuy
0.03 Faulty
0.97 Not faulty

0.15 Amart
0.06 Faulty
0.94 Not faulty

b 0.9655
14 a C and T
b $P(C \text{ and } B) = 0.34$; $P(C) \times P(B) = 0.32$ so the events are not independent
15 a 0.375
b 0.125
c 0.125
16 a 0.0278
b 0.8929
c 0.0140

17 a

x	$P(X = x)$
1	0.0278
2	0.0833
3	0.1389
4	0.1944
5	0.2500
6	0.3056

b 0.5833
18 a $P(X = x) = 0.2$
b

y	1	2	3	4
$P(Y = y)$	0.6	0.24	0.096	0.064

c 0.16
19 a 0.101 **b** 0.448 **c** 0.929 **d** 0.339
20 a $0 \leqslant X \leqslant 5$ and $19 \leqslant X \leqslant 40$
b 0.0234
21 a $X \leqslant B(10, 0.75)$
where X is the random variable 'number of patients who recover when treated'.
b 0.146
c H_0: $p = 0.75$ H_1: $p < 0.75$. $0.2142 > 0.05$ so there is insufficient evidence to reject H_0.
d 9
22 a H_0: $p = 0.3$, H_1: $p > 0.3$
b $18 \leqslant X \leqslant 40$
c 3.2%
d Reject the null hypothesis. Dhiriti's claim is supported.

Challenge
1 $x = 4$, $y = 6$, $z = 14$
2 a $X \leqslant 16$ (probability = 0.1263)
b 0.0160

CHAPTER 8

Prior knowledge 8

1 a $x = 4$ or $x = \frac{1}{5}$ **b** $x = \frac{3}{2}$ or $x = -\frac{7}{3}$
c $x = 2.26$ or $x = -0.591$ **d** $x = \pm\frac{3}{2}$
2 a $x = 10.3$, $y = 61.0°$ **b** $x = 14.8$, $y = 8.7$
3 a $833\,\text{cm}\,\text{s}^{-1}$ **b** $5000\,\text{kg}\,\text{m}^{-3}$
4 a 7.65×10^6 **b** 3.806×10^{-3}

Exercise 8A

1 a i $h = 0$ **ii** $h = 6\,\text{m}$
b $h = -48\,\text{m}$.
c Model is not valid when $x = 200$ as height would be 48 m below ground level.
2 a 90 m **b i** $h = 90\,\text{m}$ **ii** $h = 40\,\text{m}$
c $h = -1610\,\text{m}$
d Model is not valid when $t = 20$ as height would be 1610 m below sea level.
3 a $x = 2.30\,\text{m}$ or $8.70\,\text{m}$ **b** $k = 10\,\text{m}$
c When $k = 10$ metres the ball passes through the net so model not valid for $k > 10$
4 a 1320 m
b Model is valid for $0 \leqslant t \leqslant 10$
5 $0 \leqslant x \leqslant 120$
6 $0 \leqslant t \leqslant 6$

Exercise 8B

1 a Ignore the rotational effect of any external forces that are acting on it, and the effects of air resistance.
  b Ignore the frictional effects on the football due to air resistance.
2 a Ignore the rotational effect of any external forces that are acting on it, and the effects of air resistance.
  b Ignore any friction between the ice puck and the ice surface.
3 Parachute jumper and parachute should be considered together as one particle as they move together.
4 a If modelled as a light rod, the fishing rod is considered to have no thickness and is rigid.
  b If the fishing rod had no thickness and was rigid it would be unsuitable for fishing.
5 a Model golf ball as a particle, ignore the effects of air resistance.
  b Model child on sledge as a particle, consider the hill as smooth.
  c Model objects as particles, string as light and inextensible, pulley as smooth.
  d Model suitcase and handle as a particle, path as smooth, ignore friction.

Exercise 8C

1 a $18.1\,\mathrm{m\,s^{-1}}$ b $150\,\mathrm{kg\,m^{-2}}$ c $5 \times 10^{-3}\,\mathrm{m\,s^{-1}}$
  d $0.024\,\mathrm{kg\,m^{-3}}$ e $45\,\mathrm{kg\,m^{-3}}$ f $63\,\mathrm{kg\,m^{-2}}$
2 a A: Normal reaction, B: Forward thrust, C: Weight, D: Friction.
  b A: Buoyancy, B: Forward thrust, C: Weight, D: Water resistance or drag.
  c A: Normal reaction, B: Friction, C: Weight, D: Tension.
  d A: Normal reaction, B: Weight, C: Friction.

Exercise 8D

1 a $2.1\,\mathrm{m\,s^{-1}}$ b $500\,\mathrm{m}$ c $-1.8\,\mathrm{m\,s^{-1}}$
  d $-2.7\,\mathrm{m\,s^{-1}}$ e $-750\,\mathrm{m}$ f $2.5\,\mathrm{m\,s^{-1}}$
2 a $15.6\,\mathrm{m\,s^{-1}}$ b $39.8°$
3 a $5\,\mathrm{m\,s^{-2}}$ b $143°$
4 a $15.3\,\mathrm{m}$ b $24.3\,\mathrm{m}$ c $78.7°$

Mixed exercise 8

1 a $3.6\,\mathrm{m}$ b 1 m and 7 m
  c $0 \leqslant x \leqslant 8$ d $4.8\,\mathrm{m}$
2 a $7.68\,\mathrm{m}$ b $4.15\,\mathrm{m}$
  c Ignore the effects of air resistance on the diver and rotational effects of external forces.
  d Assumption not valid, diver experiences drag and buoyancy in the water.
3 a Model the man on skis as a particle – ignore the rotational effect of any forces that are acting on the man as well as any effects due to air resistance. Consider the snow-covered slope as smooth – assume there is no friction between the skis and the snow-covered slope.
  b Model the yo-yo as a particle – ignore the rotational effect of any forces that are acting on the yo-yo as well as any effects due to air resistance. Consider the string as light and inextensible – ignore the weight of the string and assume it does not stretch.

Model the yo-yo as smooth – assume there is no friction between the yo-yo and the string.
4 a $41.7\,\mathrm{m\,s^{-1}}$
  b $6000\,\mathrm{kg\,m^{-2}}$
  c $1.2 \times 10^{6}\,\mathrm{kg\,m^{-3}}$
5 a Model ball as a particle. Assume the floor is smooth.
  b i Positive – the positive direction is defined as the direction in which the ball is travelling.
    ii Negative – the ball will be slowing down.
6 a Velocity is positive, displacement is positive
  b Velocity is negative, displacement is positive
  c Velocity is negative, displacement is negative
7 a $0.158\,\mathrm{ms^{-2}}$ b $108.4°$
8 a $4.3\,\mathrm{ms^{-1}}$ b $125.5°$
9 a $158.1\,\mathrm{m}$ b $186.4\,\mathrm{m}$ c $51.3°$

CHAPTER 9

Prior knowledge 9

1 a i 3   ii 73.5
  b i 2   ii 150
  c i -1.5 ii 26.25
2 26.25 miles
3 a $x = 2$, $y = -1.5$ b $x = 1.27$ or $x = -2.77$

Exercise 9A

1 a A $80\,\mathrm{km\,h^{-1}}$, B $40\,\mathrm{km\,h^{-1}}$, C $0\,\mathrm{km\,h^{-1}}$, D $\mathrm{km\,h^{-1}}$, E $-66.7\,\mathrm{km\,h^{-1}}$
  b $0\,\mathrm{km\,h^{-1}}$ c $50\,\mathrm{km\,h^{-1}}$
2 a $187.5\,\mathrm{km}$ b $50\,\mathrm{km\,h^{-1}}$
3 a $12\,\mathrm{km\,h^{-1}}$ b 12:45
  c $-10\,\mathrm{km\,h^{-1}}$, $3\,\mathrm{km\,h^{-1}}$ d $7.5\,\mathrm{km\,h^{-1}}$
4 a $2.5\,\mathrm{m}$, $0.75\,\mathrm{s}$
  b $0\,\mathrm{m\,s^{-1}}$
  c i The velocity of the ball is positive (upwards). The ball is decelerating until it reaches 0 at the highest point.
    ii The velocity of the ball is negative (downwards), and the ball is accelerating.

Exercise 9B

1 a $2.25\,\mathrm{m\,s^{-2}}$ b $90\,\mathrm{m}$
2 a

  b $360\,\mathrm{m}$
3 a $0.4\,\mathrm{m\,s^{-2}}$ b $\frac{8}{15}\,\mathrm{m\,s^{-2}}$ or $0.53\,\mathrm{m\,s^{-2}}$ c $460\,\mathrm{m}$
4 a

  b $100\,\mathrm{s}$

Online Full worked solutions are available in SolutionBank.

5 a

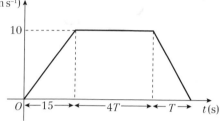

b $T = 320$ **c** $3840\,\text{m}$

6 a

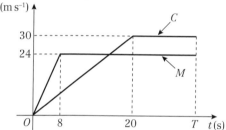

b $60\,\text{s}$

7 a $u = \frac{10}{3}$ **b** $\frac{20}{9}\,\text{m s}^{-2} = 2.22\,\text{m s}^{-2}$

8 a

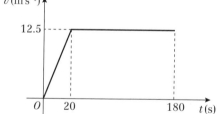

b $720\,\text{m}$

Challenge
a $6\,\text{s}$ **b** $16.5\,\text{m}$
c **i** $10.5\,\text{m}$ **ii** $4.5\,\text{m}$

Exercise 9C
1 $20\,\text{m s}^{-1}$
2 $0.625\,\text{m s}^{-2}$
3 $20\,\text{m s}^{-1}$
4 a $9\,\text{m s}^{-1}$ **b** $72\,\text{m}$
5 a $3\,\text{m s}^{-1}$ **b** $\frac{1}{3}\,\text{m s}^{-2}$
6 a $9.2\,\text{m s}^{-1}$ **b** $33.6\,\text{m}$
7 a $18\,\text{km h}^{-1}$ **b** $312.5\,\text{m}$
8 a $8\,\text{s}$ **b** $128\,\text{m}$
9 a $0.4\,\text{m s}^{-2}$ **b** $320\,\text{m}$
10 a $0.25\,\text{m s}^{-2}$ **b** $16\,\text{s}$ **c** $234\,\text{m}$
11 a $19\,\text{m s}^{-1}$ **b** $2.4\,\text{m s}^{-2}$ **c** $430\,\text{m}$
12 a $x = 0.25$ **b** $150\,\text{m}$
13 b $500\,\text{m}$

Challenge
a $t = 3$ **b** $12\,\text{m}$

Exercise 9D
1 $7\,\text{m s}^{-1}$
2 $\frac{2}{3}\,\text{m s}^{-2}$
3 $2\,\text{m s}^{-2}$
4 $0.175\,\text{m s}^{-2}$
5 a $2.5\,\text{m s}^{-2}$ **b** $4.8\,\text{s}$
6 a $3.5\,\text{m s}^{-1}$ **b** $15.5\,\text{m s}^{-1}$
7 a $54\,\text{m}$ **b** $6\,\text{s}$
8 a $90\,\text{m}$ **b** $8.49\,\text{m s}^{-1}$ (3 s.f.)

9 a $3.3\,\text{s}$ (1 d.p.)   **b** $16.2\,\text{m s}^{-1}$ (1 d.p.)
10 a $t = 4$ or $t = 8$
b $t = 4$: $4\,\text{m s}^{-1}$ in direction $\overrightarrow{AB}$, $t = 8$: $4\,\text{m s}^{-1}$ in direction $\overrightarrow{BA}$.
11 a $t = 0.8$ or $t = 4$
b $15.0\,\text{m s}^{-1}$ (3 s.f.)
12 a $2\,\text{s}$ **b** $4\,\text{m}$
13 a $0.34\,\text{m s}^{-1}$ **b** $25.5\,\text{s}$ (3 s.f.)
14 a P: $(4t + t^2)\,\text{m}$ Q: $[3(t-1) + 1.8(t-1)^2]\,\text{m}$
b $t = 6$ **c** $60\,\text{m}$
15 a $4.21\,\text{km h}^{-2}$ **b** $0.295\,\text{km h}^{-1}$

Exercise 9E
1 a $2.4\,\text{s}$ **b** $23.4\,\text{m s}^{-1}$
2 $4.1\,\text{s}$ (2 s.f.)
3 $41\,\text{m}$ (2 s.f.)
4 a $29\,\text{m}$ (2 s.f.)   **b** $2.4\,\text{s}$ (2 s.f.)
5 a $5.5\,\text{m s}^{-1}$ (2 s.f.)   **b** $20\,\text{m s}^{-1}$ (2 s.f.)
6 a $40\,\text{m s}^{-1}$ (2 s.f.)   **b** $3.7\,\text{s}$ (2 s.f.)
7 a $39\,\text{m s}^{-1}$ **b** $78\,\text{m}$ (2 s.f.)
8 $4.7\,\text{m}$ (2 s.f.)
9 a $3.4\,\text{s}$ (2 s.f.)   **b** $29\,\text{m}$ (2 s.f.)
10 $2.8\,\text{s}$ (2 s.f.)
11 a $u = 29$ (2 s.f.)   **b** $6\,\text{s}$
12 $30\,\text{m}$ (2 s.f.)
13 a $5.6\,\text{m s}^{-1}$ (2 s.f.)   **b** $3.2\,\text{m}$ (2 s.f.)

Challenge
1 a $1.4\,\text{s}$ (2 s.f.)   **b** $7.2\,\text{m}$ (2 s.f.)
2 $155\,\text{m}$ (3 s.f.)

Mixed exercise 9
1 a

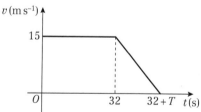

b $2125\,\text{m}$
2 a

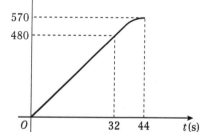

b $T = 12$
c

570
480
32 44

225

3 a i a = gradient of line. Using the formula for the gradient of a line, $a = \dfrac{v - u}{t}$, which can be rearranged to give $v = u + at$

ii s = area under the graph. Using the formula for the area of a trapezium, $s = \left(\dfrac{u + v}{2}\right)t$

b i Substitute $t = \dfrac{v - u}{a}$ into $s = \left(\dfrac{u + v}{2}\right)t$

ii Substitute $v = u + at$ into $s = \left(\dfrac{u + v}{2}\right)t$

iii Substitute $u = v - at$ into $s = \left(\dfrac{u + v}{2}\right)t$

4 $u = 8$

5 $0.165\,\text{m s}^{-2}$ (3 d.p.)

6 a $60\,\text{m}$ **b** $100\,\text{m}$

7 $1.9\,\text{s}$

8 a $4.1\,\text{s}$ (2 s.f.)   **b** $40\,\text{m s}^{-1}$ (2 s.f.)

  c air resistance

9 a $u = 11$ **b** $22\,\text{m}$

10 a $28\,\text{m s}^{-1}$ **b** $208\,\text{m}$

11 a $8\,\text{m s}^{-1}$ **b** $1.25\,\text{m s}^{-2}$ **c** $204.8\,\text{m}$

12 a $33\,\text{m s}^{-1}$ (2 s.f.)   **b** $3.4\,\text{s}$ (2 s.f.)

  c

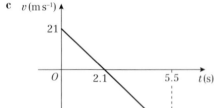

13 a $50\,\text{s}$ **b** $24.2\,\text{m s}^{-1}$ (3 s.f.)

14 $h = 39$ (2 s.f.)

15 a $32\,\text{m s}^{-1}$ **b** $90\,\text{m}$ **c** $5\,\text{s}$

16 a

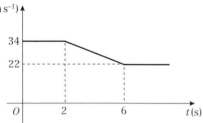

  b $180\,\text{m}$

17 a

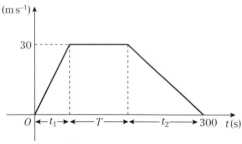

  b $\dfrac{30}{t_1} = 3x \Rightarrow t_1 = \dfrac{1}{x}, \dfrac{-30}{t_2} = -x \Rightarrow t_2 = \dfrac{30}{x}$

    So $\dfrac{10}{x} + T + \dfrac{30}{x} = 300 \Rightarrow \dfrac{40}{x} + T = 300$

  c $T = 100, x = 0.2$ **d** $3\,\text{km}$ **e** $125\,\text{s}$

Challenge

$1.2\,\text{s}$

CHAPTER 10

Prior knowledge 10

1 a $5\mathbf{i} - 3\mathbf{j}$ **b** $-4\mathbf{i} + 4\mathbf{j}$

2 a $19.2\,\text{cm}$ **b** $38.7°$

3 a $18\,\text{m s}^{-1}$ **b** $162\,\text{m}$

Exercise 10A

1 **2**

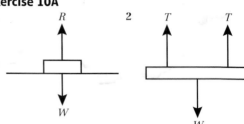

3 **4**   **5**

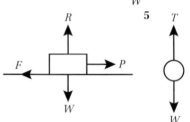

6 Although its speed is constant, the satellite is continuously changing direction. This means that the velocity changes. Therefore, there must be a resultant force on the satellite.

7 $5\,\text{N}$

8 a $10\,\text{N}$ **b** $30\,\text{N}$ **c** $20\,\text{N}$

9 a $200\,\text{N}$

  b The platform accelerates towards the ground.

10 $p = 50, q = 40$

11 $P = 10\,\text{N}, Q = 5\,\text{N}$

12 a i $20\,\text{N}$ **ii** vertically upwards

  b i $20\,\text{N}$ **ii** to the right

13 a

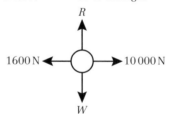

  b $8400\,\text{N}$

14 a

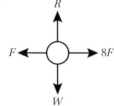

  b $600\,\text{N}$

Exercise 10B

1 a $(3\mathbf{i} + 2\mathbf{j})\,\text{N}$ **b** $\begin{pmatrix} 2 \\ -3 \end{pmatrix}\,\text{N}$

  c $(4\mathbf{i} - 3\mathbf{j})\,\text{N}$ **d** $\begin{pmatrix} 3 \\ -3 \end{pmatrix}\,\text{N}$

2 a $\mathbf{i} - 8\mathbf{j}$ **b** $-5\mathbf{i} + \mathbf{j}$

3 $a = 3, b = 4$

Online Full worked solutions are available in SolutionBank.

4 a i 5 N   **ii** 53.1°
  b i $\sqrt{26}$ N   **ii** 11.3°
  c i $\sqrt{13}$ N   **ii** 123.7°
  d i $\sqrt{2}$ N   **ii** 135°
5 a i $(2\mathbf{i} - \mathbf{j})$ N   **ii** $\sqrt{5}$ N   **iii** 116.6°
  b i $(3\mathbf{i} + 4\mathbf{j})$ N   **ii** 5 N   **iii** 036.9°
6 $a = 3, b = 1$
7 $a = 3, b = -1$
8 a $p = 2, q = -6$ **b** $\sqrt{40}$ N   **c** 18°
9 a 63.4°   **b** 3.5
10 a $a = 3, b = 2$ **b i** $\sqrt{65}$ N   **ii** 30°

Challenge
$a = 17.3$ (3 s.f.), magnitude of resultant force = 20 N

Exercise 10C

1 0.3 m s⁻² → $0.3\,\mathrm{m\,s^{-2}}$
2 39.2 N
3 25 kg
4 $1.6\,\mathrm{m\,s^{-2}}$
5 a 25.6 N   **b** 41.2 N
6 a 2.1 kg (2 s.f.)   **b** 1.7 kg (2 s.f.)
7 a $5.8\,\mathrm{m\,s^{-2}}$ **b** $2.7\,\mathrm{m\,s^{-2}}$
8 4 N
9 a $0.9\,\mathrm{m\,s^{-2}}$ **b** 7120 N   **c** 8560 N
10 a $0.5\,\mathrm{m\,s^{-2}}$ **b** 45 N
11 a 32 s   **b** 256 m
  c The resistive force is unlikely to be constant.

Challenge
a 2.9 m (2 s.f.)   **b** $3.6\,\mathrm{m\,s^{-1}}$ (2 s.f.)
c 2.16 s (3 s.f.)

Exercise 10D

1 a $(0.5\mathbf{i} + 2\mathbf{j})\,\mathrm{m\,s^{-2}}$
  b $2.06\,\mathrm{m\,s^{-2}}$ (3 s.f.) on a bearing of 014°
    (to the nearest degree).
2 0.2 kg
3 a $(21\mathbf{i} - 9\mathbf{j})$ N
  b 22.8 N (3 s.f.) on a bearing of 113°
    (to the nearest degree).
4 a $(-4\mathbf{i} + 32\mathbf{j})\,\mathrm{m\,s^{-2}}$ **b** $\left(\frac{5}{6}\mathbf{i} - \frac{1}{6}\mathbf{j}\right)\mathrm{m\,s^{-2}}$
  c $\left(-\mathbf{i} - \frac{2}{3}\mathbf{j}\right)\mathrm{m\,s^{-2}}$ **d** $\left(-\frac{4}{3}\mathbf{i} + 6\mathbf{j}\right)\mathrm{m\,s^{-2}}$
5 a $\sqrt{0.8125}\,\mathrm{m\,s^{-2}}$ on a bearing of 146°
    (to the nearest degree).
  b 6.66 s
6 $\mathbf{R} = (-k\mathbf{i} + 4k\mathbf{j})$ N
  So $4k = 3 + q$ (1), $-k = 2 + p$ (2) and $-4k = 8 + 4p$ (3)
  Adding equations (1) and (3) gives $4p + q + 11 = 0$
7 a $b = 6$ **b** $6\sqrt{2}$ N
  c $\dfrac{3\sqrt{2}}{2}\,\mathrm{m\,s^{-2}}$ **d** $\dfrac{75\sqrt{2}}{4}$ m
8 a $p = 2, q = -6$ **b** $\dfrac{25\sqrt{2}}{6}$ kg
9 0.86 kg
10 a $5 + q = -2k$ (1), $2 + p = k$ (2) and $4 + 2p = 2k$ (3)
    Adding equations (1) and (3) gives $2p + q + 9 = 0$
  b 0.2 kg

Challenge
$k = 8$

Exercise 10E

1 a 4 N
  b 0.8 N

---

  c Light ⇒ tension is the same throughout the length of the string and the mass of the string does not need to be considered. Inextensible ⇒ acceleration of masses is the same.
2 a 10 kg   **b** 40 N
3 a $2\,\mathrm{m\,s^{-2}}$ **b** 14 N
4 a 16 000 N
  b i 880 N upwards
    ii 2400 N downwards
5 a 1800 kg and 5400 kg
  b 37 000 N
  c Light ⇒ tension is the same throughout the length of the tow-bar and the mass of the tow-bar does not need to be considered. Inextensible ⇒ acceleration of lorry and trailer is the same.
6 a $2.2\,\mathrm{m\,s^{-2}}$ **b** 60 N
7 a 4 kg   **b** 47.2 N
8 a 6000 N
  b For engine, $F = ma = 3200$ N
    $R(\rightarrow)$ $12\,000 - 6000 - T = 3200$, $T = 2800$ N
9 a $R(\rightarrow)$ $1200 - 100 - 200 = 900$ N
    $F = ma$, so $a = 900 \div (300 + 900) = 0.75\,\mathrm{m\,s^{-2}}$
  b 325 N   **c** 500 N

Exercise 10F

1 a 33.6 N (3 s.f.)
  b $2.37\,\mathrm{m\,s^{-1}}$ (3 s.f.)
  c 2.29 m (3 s.f.)
2 a $2mg$ N
  b For P: $2mg - kmg = \frac{1}{3}kmg$
    So $2 - k = \frac{1}{3}k$ and $k = 1.5$
  c Smooth ⇒ no friction so magnitude of acceleration is the same in objects connected by a taut inextensible string.
  d While Q is descending, distance travelled by $P = s_1$
    $s = ut + \frac{1}{2}at^2 \Rightarrow s_1 = \frac{1}{6}g \times 1.8^2 = 0.54g$
    Speed of P at this time $= v_1$
    $v^2 = u^2 + 2as \Rightarrow v_1^2 = 0^2 + \left(2 \times \dfrac{g}{3} \times 0.54g\right) = 0.36g^2$
    After Q hits the ground, P travels freely under gravity and travels a further distance s_2.
    $v^2 = u^2 + 2as \Rightarrow 0^2 = 0.36g^2 - 2gs_2 \Rightarrow s_2 = 0.18g$
    Total distance travelled $= s_1 + s_2 = 0.54g + 0.18g$
    $= 0.72g$ m
    As particles started at same height P must be s_1 metres above the plane at the start.
    Maximum height reached by P above the plane $= 0.72g + s_1 = 0.72g + 0.54g = 1.26g$ m
3 a $s = ut + \frac{1}{2}at^2$ so $2.5 = 0 + \frac{1}{2} \times a \times 1.25^2$, $a = 3.2\,\mathrm{m\,s^{-2}}$
  b 39 N
  c For A, $R(\downarrow)$: $mg - T = ma$
    $T = m(9.8 - 3.2)$, $T = 6.6m$
    Substituting for T: $39 = 6.6m$
    $m = \frac{65}{11}$
  d Same tension in string either side of the pulley.
  e $\frac{40}{49}$ s
4 a $0.613\,\mathrm{m\,s^{-2}}$ (3 s.f.)
  b 27.6 N (3 s.f.)
  c 39.0 N (3 s.f.)
5 a i $2.84\,\mathrm{m\,s^{-2}}$ (3 s.f.)
    ii $2.84(1.5) = 1.5g - T$
      $T = 1.5g - 4.26 = 10.4$ N (3 s.f.)
    iii 3.3 N
  b Same tension in string either side of the pulley.

Mixed exercise 10

1 a

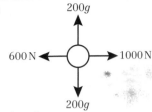

200g

600 N ← ○ → 1000 N

200g

b $2\,\mathrm{m\,s^{-2}}$

2 1000 N (2 s.f.) vertically downwards

3 a 2000 N   **b** 36 m

4 a $1.25\,\mathrm{m\,s^{-2}}$ **b** 6 N

5 Res($\rightarrow$) $3R - R = 1200 \times 2 \Rightarrow R = 1200$
Driving force $= 3R = 3600\,\mathrm{N}$

6 $(28\mathbf{i} + 4\mathbf{j})\,\mathrm{m\,s^{-2}}$

7 $a = 1, b = -3$

8 a $\sqrt{5}\,\mathrm{m\,s^{-2}}$ **b** $\dfrac{9\sqrt{5}}{2}\,\mathrm{m}$

9 a $a = -15, b = 12$
b **i** $11.7\,\mathrm{m\,s^{-2}}$ (3 s.f.) on a bearing of $039.8°$ (3 s.f.)
ii $52.7\,\mathrm{m}$ (3 s.f.)

10 a $0.7\,\mathrm{m\,s^{-2}}$ **b** 770 N   **c** 58 m
d Inextensible $\Rightarrow$ tension the same throughout, and the acceleration of the car and the trailer is the same.

11 a R($\rightarrow$) $8000 - 500 - R = 3600 \times 1.75, R = 1200\,\mathrm{N}$
b 2425 N   **c** 630 N (2 s.f.)

12 a $\frac{1}{3}g\,\mathrm{m\,s^{-2}}$ **b** $3.6\,\mathrm{m\,s^{-1}}$ **c** $2\frac{2}{3}\,\mathrm{m}$
d **i** Acceleration both masses equal.
ii Same tension in string either side of the pulley.

13 a $\frac{12}{7}g\,\mathrm{N}$ **b** $m = 1.2$

14 a $3.2\,\mathrm{m\,s^{-2}}$ **b** 5.3 N (2 s.f.)   **c** $F = 3.7$ (2 s.f.)
d The information that the string is inextensible has been use in part **c** when the acceleration of A has been taken to be equal to the acceleration of B.

15 a **i** $0.5g - T = 0.5a$ **ii** $T - 0.4g = 0.4a$
b $\frac{4}{9}g\,\mathrm{N}$ **c** $\frac{1}{9}g\,\mathrm{m\,s^{-2}}$ **d** 0.66 s (2 s.f.)

Challenge
$k = -\frac{5}{2}$

CHAPTER 11

Prior knowledge 11

1 a $6x - 5$ **b** $\dfrac{x^{-(\frac{1}{2})} - 12}{(x^3)}$

2 a $(1.5, -4.75)$ **b** $(1, 9)$ and $(3, 5)$

3 a $\dfrac{5x^2}{2} + 8x + 1$ **b** $x^3 - x^2 + 5x + 7$

4 a 18   **b** $11\frac{1}{3}$

Exercise 11A

1 a 8 m   **b** $t = 0$ and $t = \pm 3$
2 a 4 m   **b** 6 m
3 a $7\,\mathrm{m\,s^{-1}}$ **b** $9.25\,\mathrm{m\,s^{-1}}$
c $-11\,\mathrm{m\,s^{-1}}$; body is travelling in opposite direction.
4 a 0.8 m   **b** 4 s
c 1.6 m   **d** $0 \le t \le 4$
5 a $8\,\mathrm{m\,s^{-1}}$ **b** $t = \frac{4}{3}$ and $t = 2$
c $t = \frac{1}{3}$ and $t = 3$ **d** $8\,\mathrm{m\,s^{-1}}$
6 a 4 s   **b** $8\,\mathrm{m\,s^{-1}}$
7 $T = 3$: returns to starting point and $s = 0$ when $t = 0$ and $t = 3$.
8 a $t = \frac{1}{3}$ and $t = 3$ **b** $\frac{16}{15}\,\mathrm{m\,s^{-1}}$

Exercise 11B

1 a **i** $v = 16t^3 + \dfrac{1}{t^2}$ **ii** $a = 48t^2 - \dfrac{2}{t^3}$

b **i** $v = 2t^2 - \dfrac{2}{t^3}$ **ii** $a = 4t + \dfrac{6}{t^4}$

c **i** $v = 18t^2 + 30t - 2$ **ii** $a = 36t + 30$

d **i** $v = \dfrac{9t^2}{2} - 2t - \dfrac{5}{2t^2}$ **ii** $a = 9t - 2 + \dfrac{5}{t^3}$

2 a $46\,\mathrm{m\,s^{-1}}$ **b** $24\,\mathrm{m\,s^{-2}}$
3 $7\,\mathrm{m\,s^{-2}}$ in the direction of x decreasing.
4 6.75 m
5 a $k = 4$ **b** $a = -4\,\mathrm{m\,s^{-2}}$
6 1.7 cm

Exercise 11C

1 a 0.25 s
b 4.54 m
c $v = -1.88\,\mathrm{m\,s^{-1}}$
2 a The body returns to its starting position 4 s after leaving it.
b Since $t \ge 0$, t^3 is always positive.
Since $t \le 4$, $4 - t$ is always non-negative.
c 27 m
3 a

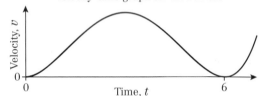

Velocity–time graph for motion of P

b $v = 81\,\mathrm{m\,s^{-1}}$ when $t = 3\,\mathrm{s}$
4 a Discriminant of $2t^2 - 3t + 5$ is <0, so no solutions for $v = 0$
b $3.88\,\mathrm{m\,s^{-1}}$ (3 s.f.)
5 a

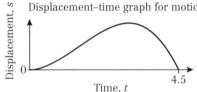

Displacement–time graph for motion of P

b s is a distance so cannot be negative.
c 13.5 m   **d** $9\,\mathrm{m\,s^{-2}}$
6 Max distance is when $\dfrac{ds}{dt} = 3.6 + 3.52t - 0.06t^2 = 0$,
so $t = 59.7$ (3 s.f.)
$\therefore$ Max distance $= 2.23\,\mathrm{km}$ (3 s.f.), so the train never reaches the end of the track.

Exercise 11D

1 a $s = t^3 - t$ **b** $s = \dfrac{t^4}{2} - \dfrac{t^3}{2}$ **c** $s = \dfrac{4}{3}t^{\frac{3}{2}} + \dfrac{4t^3}{3}$

2 a $v = 4t^2 - \dfrac{2t^3}{3}$ **b** $v = 6t + \dfrac{t^3}{9}$

3 12 m
4 a $v = 6 + 16t - t^2$ **b** -6
5 42.9 m (3 s.f.)
6 12.375 m
7 a $10\frac{2}{3}$ **b** 13 m
8 $t = \frac{3}{2}$ and $t = 5$
9 a $t = 1$ and $t = 5$ **b** 6 m

10 $T = 1.5\,\text{s}$

11 a $v = \dfrac{t^2}{2} - 3t + 4$ **b** $t = 2$ and $t = 4$ **c** $\dfrac{2}{3}\,\text{m}$

12 a $86\,\text{m}$ **b** $60\,\text{m s}^{-1}$

Challenge

$\dfrac{200}{3}\,\text{m}$

Exercise 11E

1 $v = \int a\,\mathrm{d}t = at + c$

$a \times 0 + c = 0 \Rightarrow c = 0 \Rightarrow v = at$

$s = \int v\,\mathrm{d}t = \int at\,\mathrm{d}t = \tfrac{1}{2}at^2 + k$

$\tfrac{1}{2}a \times 0^2 + k = x \Rightarrow k = x$

so $s = \tfrac{1}{2}at^2 + x$

2 a $a = 5$, $v = \int 5\,\mathrm{d}t = 5t + c$; when $t = 0$, $u = 12$ so
$c = 12$, $v = 12 + 5t$

b $s = \int 12 + 5t\,\mathrm{d}t = 12t + \dfrac{5t^2}{2} + d$, when $t = 0$,

$s = 7$ so $d = 7$, $s = 12t + 2.5t^2 + 7$

3 $v = \dfrac{\mathrm{d}s}{\mathrm{d}t} = u + at$; $\dfrac{\mathrm{d}v}{\mathrm{d}t} = a$ so constant acceleration a

4 A $a = 4 - 6t$, not constant

B $a = 0$. no acceleration

C $a = \tfrac{1}{2}$, constant

D $a = -\dfrac{12}{t^4}$ not constant

E $v = 0$, particle stationary

5 a $4\,\text{m s}^{-2}$ **b** $p = 2, q = 5, r = 0$

6 a $680\,\text{m}$

b $\dfrac{\mathrm{d}s}{\mathrm{d}t} = 25 - 0.4t \Rightarrow \dfrac{\mathrm{d}^2s}{\mathrm{d}t^2} = -0.4$ ∴ a is constant

c $420\,\text{m from } A$

Mixed exercise 11

1 a $t = 5$

b $37.5\,\text{m}$

2 a $30\,\text{m s}^{-2}$

b $75\,\text{m}$

3 a Displacement $= 8t + t^2 - \dfrac{t^3}{3}$

b Max displacement when $t = 4$, $s = 26\tfrac{2}{3}\,\text{m}$, which is less than $30\,\text{m}$ so P does not reach B.

c $t = 6.62\,\text{s}$

d $53\tfrac{1}{3}\,\text{m}$

4 a $\dfrac{32}{3}\,\text{m s}^{-1}$ **b** $\dfrac{40}{3}\,\text{m}$

5 a $(-3t^2 + 22t - 24)\,\text{m s}^{-1}$ **b** $t = \tfrac{4}{3}$ and $t = 6$

c $t = \dfrac{11}{3}$

d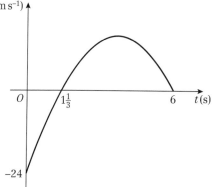

e $0 \leqslant t < 0.38, \dfrac{10}{3} < t < 4$

6 a $t = \tfrac{5}{3}$ and $t = 2$ **b** $13\,\text{m s}^{-2}$ **c** $\dfrac{433}{27}\,\text{m}$

7 a $v = \dfrac{t^4}{2} - 4t^2 + 6$ **b** $s = \dfrac{t^5}{10} - \dfrac{4t^3}{3} + 6t$

c $t = \sqrt{2}$ and $t = \sqrt{6}$

8 max $= 8.64\,\text{m}$, min $= 1.14\,\text{m}$

9 $a = 1500, b = 800, c = -16$

10 a $v = \int 20 - 6t\,\mathrm{d}t = 20t - 3t^2 + c$
At $t = 0$, $v = 7$ so $c = 7$ and $v = 7 + 20t - 3t^2$

b The greatest speed is $40\tfrac{1}{3}\,\text{m s}^{-1}$

c $196\,\text{m}$

11 $v = \int k(7 - t^2)\,\mathrm{d}t \Rightarrow v = k\left(7t - \dfrac{t^3}{3}\right) + c$

$t = 0, v = 0 ∴ c = 0; t = 3, v = 6 ∴ k = \tfrac{1}{2}$

$v = \dfrac{7}{2}t - \dfrac{t^3}{6}$

$s = \int v\,\mathrm{d}t = \int \left(\dfrac{7}{2}t - \dfrac{t^3}{6}\right)\mathrm{d}t = \dfrac{7t^2}{4} - \dfrac{t^4}{24} + c$

$t = 0, s = 0 ∴ c = 0$

$s = \dfrac{7t^2}{4} - \dfrac{t^4}{24} = \dfrac{1}{24}t^2(42 - t^2)$

12 a Time cannot be negative so $t \geqslant 0$
at $t = 5$ $s = 0$ so mouse has returned to its hole.

b $39.1\,\text{m}$

13 a Mass is not constant as fuel is used.
Gravity is not constant so weight not constant.
Thrust may not be constant.

b $v = (1.69 \times 10^{-7})\,t^4 - (1.33 \times 10^{-4})\,t^3 + 0.0525\,t^2 + 0.859\,t + 274\,\text{m s}^{-1}$

c $v = 5990\,\text{m s}^{-1}$

d 510 seconds (2 s.f.) after launch

Challenge

1 $32.75\,\text{m}$

2 $91\,\text{m s}^{-1}$

Review exercise 2

1 a Constant acceleration

b Constant speed **c** $30.5\,\text{m}$

2 a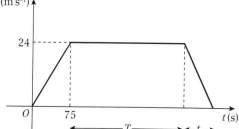

b $0.48\,\text{m s}^{-2}$ **c** $T = 250$ **d** $375\,\text{s}$

3 a $185\,\text{s}$ **b** $2480\,\text{m}$

c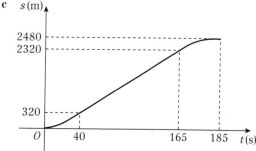

4 a $28\,\text{m s}^{-1}$ **b** $5.7\,\text{s}$ (2 s.f.)

5 $t = 2$ and $t = 4$

6 $q = \sqrt{10}$ and $p = \sqrt{30}$

7 $66\frac{2}{3}$ m

8 a 0.693 m s^{-2} (3 s.f.)   **b** 7430 N (3 s.f.)

   c **i** Rotational forces and air resistance can be ignored.

      ii The tension is the same at both ends and its mass can be ignored.

9 a Ball will momentarily be at rest 25.6 m above A.

    $0^2 = u^2 + 2 \times 9.8 \times 25.6$, $u = 22.4$

   b 4.64 (3 s.f.)   **c** 6380 (3 s.f.)

   d v (m s^{-1})

   e Consider air resistance due to motion under gravity.

10 a 4.2 m s^{-2} **b** 3.4 N (2 s.f.)

    c 2.9 m s^{-1} (2 s.f.)   **d** 0.69 s (2 s.f.)

    e **i** String has negligible weight.

       ii Tension in string is constant i.e. same at A and B.

11 a 2.9 N (2 s.f.)   **b** 4.9 m s^{-2} **c** 0.21 s (2 s.f.)

    d Same acceleration for P and Q.

12 a **i** 1050 N   **ii** 390 N

    b 3 m s^{-2}

13 a 8697 N   **b** 351 N   **c** 507 N

14 a 63°

    b $2 + \lambda = k$ (1) and $3 + \mu = 2k$ (2)

      $2 \times$ (1) = (2) so $4 + 2\lambda = 3 + \mu$ so $2\lambda - \mu + 1 = 0$

    c 4.47 (3 s.f.)

15 a 17.5 (1 d.p.)

    b 66°

    c $P = 3\mathbf{i} + 12\mathbf{j}$

      $Q = 4\mathbf{i} + 4\mathbf{j}$

16 6 s

17 a 4.5 m s^{-1} **b** 4.5 s

18 a $t = 1$ and $t = \frac{5}{3}$ **b** 16 m s^{-2} **c** 4 m

19 a $6 - 3t^{\frac{1}{2}}$ **b** $3t^2 - \frac{4}{5}t^{\frac{5}{2}}$

Challenge

1 $t_1 = 62.2$ s, $t_2 = 311.1$ s, $t_3 = 46.7$ s (3 s.f.)

  Distance = 20.6 km (3 s.f.)

2 a $a = 7.4$ m s^{-2} **b** 39 N

   c 13 N   **d** 55 N (2 s.f.)

   e Acceleration is the same for objects connected by a taut inextensible string.

Practice paper

1 a 0.15

    b $P(B) \times P(M) = 0.1575$ so the events are not independent.

2 a Continuous – measured variable can take any value

    b 14.01, 1.36 (3 s.f.)

    c Increase – value higher than current mean

    d Clare could select random days in September. She could include data from other UK locations for September 2015.

3 a 0.2

    b 0.65

    c **i** 0.1757   **ii** 0.0260

4 a Test statistic is the number of plates that are flawed.

    H_0: $p = 0.3$, H_1: $p < 0.3$

    b 0, 1, 2   **c** 3.55%

    d 1 falls into the critical region therefore there is evidence to support the claim.

5 a Increase in energy released for each degree of temperature.

    b Value of h is a long way from the range of the experimental data so it would not be sensible – extrapolation.

    c The regression line should only be used to predict a value of e given h so it would not be sensible.

6 0.87 (2 d.p.)

7 a 0.75 m s^{-2} **b** 845 N

    c Same acceleration for car and trailer

8 a $a = 3\mathbf{i} - \mathbf{j}$ m s^{-2}

    b 18.4° below

    c $\sqrt{10}$ m s^{-2}

9 a 24.5 m s^{-1} **b** 30.625 m

    c $\frac{5}{7}$ s and $\frac{30}{7}$ s

10 $\frac{153}{16}$ m

Online   Full worked solutions are available in SolutionBank.

Index